CAN SCIENCE AND TECHNOLOGY SAVE CHINA?

CAN SCIENCE AND TECHNOLOGY SAVE CHINA?

Edited by Susan Greenhalgh and Li Zhang

CORNELL UNIVERSITY PRESS   ITHACA AND LONDON

First published 2020 by Cornell University Press

Library of Congress Cataloging-in-Publication Data

Names: Greenhalgh, Susan, editor. | Zhang, Li, 1965 May- editor.
Title: Can science and technology save China? / edited by Susan Greenhalgh and Li Zhang.
Description: Ithaca [New York] : Cornell University Press, 2020. | Includes bibliographical references and index.
Identifiers: LCCN 2019017310 (print) | LCCN 2019018070 (ebook) | ISBN 9781501747045 (pdf) | ISBN 9781501747052 (epub/mobi) | ISBN 9781501747021 | ISBN 9781501747021 (cloth) | ISBN 9781501747038 (pbk.)
Subjects: LCSH: Science—Social aspects—China. | Technology—Social Aspects—China. | Science and state—China. | Technology and state—China.
Classification: LCC Q175.52.C6 (ebook) | LCC Q175.52.C6 C35 2020 (print) | DDC 338.951/06—dc23
LC record available at https://lccn.loc.gov/2019017310

Contents

Acknowledgments

This book would not have been possible without the support of numerous institutions and individuals. The project emerged from a workshop in China anthropology titled "A Better Life through Science and Biomedicine?" held at Harvard University on April 15 and 16, 2016. Big thanks go to Harvard's Fairbank Center for Chinese Studies for hosting the workshop, and to Mark Grady, the center's events coordinator, for the organizational skill, unflagging energy, and good humor he devoted to making sure everything unfolded smoothly. Several organizations on campus contributed generously to funding the event. They include the Fairbank Center, the Weatherhead Center for International Affairs, and the Department of Anthropology. At the workshop, two discussants—Joe Dumit, Professor of Anthropology and then-Director of Science and Technology Studies at the University of California, Davis, and Wen-Hua Kuo, Associate Professor, School of Humanities and Social Sciences at National Yang-Ming University (Taipei) and editor of the noted journal *East Asian Science, Technology, and Society: An International Journal*—provided thought-provoking commentaries. In addition to those whose chapters appear here, Marty (Lindsey) Alexander, Nancy N. Chen, and Anna Lora-Wainwright presented papers at the workshop. The comments and ideas they shared at that event are reflected in the essays published here. Warm thanks go to Shuang Lu Frost and Shanni Zhao, advanced graduate students in anthropology at Harvard, for their able assistance with and participation in the workshop.

All the contributors to this book are grateful to two anonymous reviewers for the press for their exceptionally smart and professionally generous comments on the book as a whole and each of the individual chapters. Special thanks are due Jim Lance, our editor at Cornell University Press, whose sharp editorial eye and finely honed negotiation skills ensured that this book was externally reviewed and formally accepted in admirably short order. Karen Laun, our production editor at the press, and Kristen Bettcher, project editor at Westchester Publishing Services, successfully guided the book through the production process with remarkable speed and efficiency. To all these individuals and institutions we owe big debts of gratitude.

—The editors

CAN SCIENCE AND TECHNOLOGY SAVE CHINA?

GOVERNING THROUGH SCIENCE

The Anthropology of Science and
Technology in Contemporary China

Susan Greenhalgh

National Rejuvenation and Ascent through Modern Science: Political Promise, Political Mandate

Since its embrace of modernity a century ago, China has been animated by official dreams of national revival and global ascent. Central to every dream has been the promise of modern science. In the early twentieth century, Western-oriented Chinese intellectuals embraced modern science with fervor, seeing the promotion of "Mr. Science" (*sai xiansheng*) as a powerful means to critique China's traditional culture, whose scientific backwardness was seen as a root of China's poverty, and to set the nation on the path to modern civilization (Kwok 1965). Since that time, science has been associated with modernity and national salvation and imbued with almost omnipotent powers. One could perhaps say that *kexue jiuguo* and, more recently, *kexue xingguo*—popular adages conveying that China can and should be saved and rejuvenated by modern science—have been built into the cultural DNA of the Chinese nation.

In China, this vital project of defending and strengthening the nation with modern science has been conceived and carried out mostly by successive states. For over a century, science has been largely an instrument of state rule, introduced through top-down initiatives directed at strategic, state-defined ends (Elman 2005; Shen and Williams 2005). And so, far from withering under the Communists, the belief in the promise and power of science persisted into the early People's Republic. Under Mao Zedong (paramount leader 1949–1976), science was assigned a highly progressive role in Marxist philosophy (Kwok 1965). In the 1950s, following the lead of the Soviet Union, the leaders of the young People's Republic consolidated state control over science and created Soviet-style research institutions. During the chaotic years of the Cultural Revolution (1966–1976), Mao notoriously turned on the scientists, wreaking havoc on the nation's science and social science establishments.

In a striking historic reversal, in the years since Mao's death, modern science and technology (below, simply S&T) have secured a political prominence perhaps unmatched in the world. After decades of often disastrous ideological decision-making culminating in the Cultural Revolution, modern science appeared as the way out, a deus ex machina that would guide China into the modern world. Deng Xiaoping, Mao's successor as paramount leader (1978–1989), designated science and technology the first of the "four modernizations" (*sihua*) and China's primary route to modernity and global power. (Chinese political discourse joins "science" and "technology" [*keji*] into a term of central importance in the post-Mao era. My discussion of the political significance of these domains follows Chinese usage and yokes the terms, except where the emphasis is clearly on either science or technology.) Starting in the late 1970s, the new reform leadership under Deng began to invest in S&T as a national strategy, rebuilding the science infrastructure and introducing a series of policies to promote the rapid development of modern science and technology. "Scientific policymaking" (*kexue juece*) became politically obligatory, and science—that is, the claim to be a scientific modernizer—joined the now-exhausted Marxism-Leninism–Mao Zedong Thought as a legitimating ideology of the Deng regime.

In the twenty-first century, science and the party-state remain deeply intertwined. Following Deng, successive leaders have adjusted the political meanings and uses of the term *science* to meet new goals, yet science has maintained its importance. Hu Jintao (top leader 2002–2012) made the "scientific concept of development" (*kexue fazhan guan*) a signature theme, to be incorporated into all policy arenas (Fewsmith 2008; Wang 2009). During the 1980s, 1990s, and early 2000s, as the country's leadership filled with scientists and engineers, China became a virtual technocracy run largely by engineer-politicians (Li 2016; also Andreas 2009). Since the late 2000s and early 2010s, technocratic dominance of the party-state has waned. Although the majority of the nation's top leaders are now trained in economics, law, politics, and the humanities, the regime continues to place immense faith in the powers of S&T (Li 2016, chap. 5).[1]

Today China aspires to be one of the most technologically innovative nations by 2020 and a global S&T powerhouse by midcentury, and it is investing heavily to realize those goals (Cao and Suttmeier 2017; Chinese Academy of Sciences [CAS] 2016). These widely promoted aspirations of the leadership have given rise to a global narrative of China as an increasingly formidable global power in science and, even more so, technology. In the Western media, high-tech success is the dominant story about China's scientific and technological development. And there is much that is impressive. A global leader in a few fields, China is now home to more researchers than any other country, and it is second only to the United States in the number of scientific publications (Yu, Zhang, and Lai 2014). Top leader Xi Jinping (from late 2012; likely to remain in power well beyond the usual ten years) has energetically supported this agenda, pouring vast sums into advanced technology projects such as Made in China 2025 and Internet+, which are aimed at spreading robotics, networking, and artificial intelligence among existing industrial sectors (Naughton 2018).[2] Addressing a major S&T conference in 2016, he stressed the role of science and technology as bedrocks on which China relies for its power, enterprises rely for their success, and ordinary people rely for a better life (CAS 2016). Xi has advocated strengthening basic research, yet he has also asked for translatable results that will help solve enduring economic and industrial problems ("The Future of Chinese Research," 2016). Reflecting the leadership's view that technological development is the essential key to making China globally competitive and addressing the nation's confounding domestic problems, the latest five-year development plan (the thirteenth, covering 2016–2020) prioritizes innovation in S&T (Cyranoski 2016; Five-Year Plan 2016; Yu, Zhang, and Lai 2014). Forty years after Deng elevated science and technology to the top of the list of domains to be modernized, the official narrative of S&T as China's domestic savior and global uplifter is stronger than ever.

The official narrative, however, tells only part of the story. The post-Mao years have brought the rapid development not only of science, but also of *scientism*, the belief in science as a panacea for all the nation's ills. Indeed, scientism and its twin, *technicism*, which values instrumental reasoning and technical efficiency above all, emerged as virtual official ideologies of the party-state (Greenhalgh 2008; Hua 1995; Suttmeier 1989). In the West, the years following World War II saw the emergence of widespread critiques of the adverse effects of powerful technologies and the dominance of technological rationality in modern society. In China, in sharp contrast, modern science and technology have been surrounded by a halo of official optimism and largely immune to social critique (Shen and Williams 2005). Post-Mao China has been home to a veritable state-sponsored religion of S&T marked by a widespread faith in the power of modern science and technology to solve the problems that other approaches have failed to solve. Since the early 1980s, the vision of mobilizing science to remake China has deeply penetrated Chinese society, reshaping the mindset of ordinary people. In the late twentieth century, the official scientism and technicism of China's leaders increasingly became a mass culture of S&T, in which modern science, statistics, and technology were at times treated with almost magical or mystical reverence, their products accepted, with few questions, as ideal solutions to China's problems. In the twenty-first century, despite widespread complaints about the vexing problems of daily life—contaminated food and toxic air, for example—there has been great pride in the nation's high-tech achievements and little apparent discussion of the potential dangers of a state-S&T-driven modernization project in which society has no independent voice. Only in the last decade or so has that begun to change, as some high-tech fixes to the nation's environmental crises have failed to work, and popular discontent has become hard to ignore (Economy 2018, 152–185). Patient attacks on physicians, which have grown violent in recent years, represent an assault on scientific as well as clinical authority and evidence of the limits of technological solutions to the problems plaguing China's healthcare system (Nie et al. 2018).

Today, under Xi Jinping, the party's promise of a brighter future takes the form of the China Dream (*zhongguo meng*). Introduced by Xi in late 2012, the phrase is now widely used in official statements and has become one of the guiding ideological principles of the leadership under Xi. A combination ideological campaign and nation-building project to transform China into a global economic and military power, the China Dream aims to connect the party to the people through a common vision by addressing social inequalities, restoring Confucian values, and fostering a sense of personal well-being (Carlson 2015; a darker view is presented in Economy 2018). Xi's signature slogan is presented as the people's dream; indeed, cities, towns, and villages across the nation have sprouted "dream walls" plastered with images of happy Chinese extolling the virtues of socialism and their leader's ambitious plans.

On-the-Ground Realities: The Rise of an Anthropology of Chinese Science and Technology

Meantime, as if in mockery of Xi's China Dream, the on-the-ground reality of life in China today poses manifold threats to human flourishing. A large body of science and social science research makes clear that the party-state's forty-year pursuit of economic-development-at-any-cost has eroded human health and undermined the ecological balance that is necessary to sustain life. Even as infectious diseases continue to erupt unpredictably, the chronic diseases of modernity (cardiovascular and respiratory diseases, cancers, diabetes, and more) are taking an enormous toll on human vitality. Mental distress and mental illness plague untold numbers of rural and urban residents. In much of the country, severe soil, water, and air pollution poses dire threats to human well-being. After countless scandals, the safety of the food and medicine in China's markets seems always to be in question, spawning widespread feelings of vulnerability, fear, and "bio-insecurity" about the adequacy of resources for human survival (Chen 2014). In light of these troubling realities, the China Dream seems best described as an instance of what Lauren Berlant (2011) calls "cruel optimism," in which the promised hope serves to stifle dissent, discourage change, and render aspirations unattainable.

With individual, collective, and environmental health all now in crisis, the restorative powers of modern science and technology are being sought after with increased urgency. In the 2010s, however, modern sciences and technologies are being summoned to rescue China not from the depredations of an imperial West or from the blunders of Mao's collectivism, but from the fallout of post-1978 party policies that have prioritized economic growth above all else. Given that the party that produced those policies is still in power, this project at times takes on a kind of mission-impossible character. Tasked with finding solutions, Chinese experts have been adapting a wide range of international sciences to the China context, laboring to create scientific knowledge in a context in which science is always already highly politicized and in which the mantra of "economy first" can scarcely be challenged. For their part, China's people, facing threats to their psychological, physical, socioeconomic, and even moral well-being (Zhang, Kleinman, and Tu 2011), have been responding by latching onto the promise of science to restore life, embracing solutions that they believe are based on the latest sciences and technologies. The proliferation of these science projects, at both expert and mass-society levels, raises a host of questions. In the contemporary Chinese context, in which the meaning of science has been unmoored from specific knowledge-producing activities and attached to political dreams articulated by the party-state, what counts as "modern science" to China's people?[3] What hopes are being invested in it? Who is making "science" and how? Are these scientific and technological solutions having their promised effects, or are they producing other effects that may be neither anticipated nor wanted?

Given the political centrality of science and technology in contemporary China (here, meaning the reform era that began in the late 1970s), one would expect to find a large body of social science scholarship on their making, workings, and effects. In the early reform years, the 1980s and 1990s, these domains of Chinese life received remarkably little attention. Since the mid-2000s, though, scholarly interest has grown quickly.[4] Political scientists (and a few political sociologists) have taken the lead, tracing the development of the nation's post-1978 S&T policies and exploring the political roots, organization, and applications of Chinese science (major works include Baum 1980; Miller 1996; Saich 1989; Simon and

Goldman 1989; Suttmeier 1980; Wang 1993; on science-party-state relations, Hamrin and Cheek 1986; Goldman 1987, 1994; Goldman and MacFarquhar 1999; critical science-policy updates include Cao et al. 2013; Cao and Suttmeier 2017). More recent political science and sociology work has focused on technology policy and innovation, China's S&T elite and talent pool, and the role of government-sponsored think tanks in supplying social science expertise for policymaking (Cao 2004; Li 2001; Li 2017; Sigurdson 2006; Simon and Cao 2009; Suttmeier, Cao, and Simon 2006; Suttmeier and Yao 2004; see also Sleeboom-Faulkner 2007).

In the last decade, as the government has focused its energies on transforming China into an "innovation nation" and its thinking on innovation has broadened beyond science and research and development to consider the larger ecosystem (of market forces, entrepreneurship, and the financial and legal set-up, as well as state policy), the scholarship on Chinese technology has grown rapidly in scale and diversity. Today, scholars in economics and management, geography, and ur-ban and regional studies are tracing China's push to become a global innovation hub, analyzing technology development by industrial sector, the roles of the party-state and global market in those dynamics, and the complex state-university-business relationships that support (and hinder) the deepening of the country's innovation capacity (recent illustrative works include Chen and Naughton 2016; Fuller 2016; Lewin, Kenney, and Murmann 2016; Naughton 2019; Yip and Mc-Kern 2016; Zhou, Lazonick, and Sun 2016).

This burgeoning scholarship tells important parts of the story of Chinese sci-entific and technological advance. Yet by centering the party-state and focusing on institutions, existing work leaves vital questions unexplored. Virtually the en-tire domain of science and society remains untouched. So too does the cognitive core of science—the hypotheses, methods, data, and so on that form the stuff of science. How do Chinese experts create scientific knowledge and technical inno-vations in the unusual context of the People's Republic? How do non-state insti-tutions (corporations or nonprofits, for example) mobilize science in pursuit of their agendas? How do members of society at large take up, negotiate, and/or con-test the sciences and technologies offered as solutions to their problems? Why and to what extent do they matter in ordinary people's lives? The study of Chi-nese science remains underdeveloped not only in contemporary Chinese stud-ies, but also in the interdisciplinary field of science and technology studies (STS), which, though becoming broader in scope, continues to prioritize the experiences of Euro-America.[5] Even as China moves ever closer to its goal of becoming a global S&T power, the nation's distinctive approaches to developing and applying sci-ence are largely missing from the field of STS. This limited attention to Chinese science means missed opportunities for China scholars and science studies schol-ars alike.

In the last decade or so, anthropologists of contemporary China, some influenced by STS, have begun to explore precisely these questions. Ethnographic research on the sciences of psychiatry (Chen 2003; Lee 2011; Yang 2015; Zhang 2014), disability (Kohrman 2005), population/reproduction (Gottschang 2018; Greenhalgh 2008; Greenhalgh and Winckler 2005; Wahlberg 2018), traditional Chinese medicine (Palmer 2007; Zhan 2009; see also Farquhar 1996 on medical expertise), sexuality (Farquhar 2002; Zhang 2015), public health (Hyde 2007; Mason 2016), cancer epidemiology (Lora-Wainwright 2013), the environment (Hathaway 2013; Tilt 2014), genome sequencing (Ong 2016, 197–222), and regenerative medicine (Song 2017) has shown how science has profoundly restructured social, cultural, and political life in the post-Mao era, but not necessarily in the ways intended.[6] The party-state remains a key actor in engineering dreams of personal and national rejuvenation through science and technology, yet the dreams acquire social lives of their own as they get taken up and put into practice by diverse social forces. While the number of scholars with such interests is growing rapidly, anthropological research on science and society in China has not been cumulative, in the sense of creating shared conversations across domains of science or living.

In April 2016 I invited Li Zhang, the coeditor of this book, to join in convening a workshop-style conference at Harvard University to explore these questions and, in the process, create a subfield of scholars with shared concerns. Taking advantage of the growing interest in questions of environmental sustainability among younger China anthropologists, we sought to bring together researchers working in medical and environmental anthropology, subfields that until recently have developed relatively independently (newer works exploring the intersections of pollution and health include Bunkenborg 2014; Lamoreaux 2016; Wahlberg 2018). We asked participants to write about Chinese dreams of modern S&T rejuvenating the nation. Most wrote not about hopes and dreams, but about fears, nightmares, and struggles to achieve the promised good life and good society through scientific and technological means. Overall, the presentations conveyed a bleak picture of contemporary Chinese life in which the mental, bodily, and environmental costs of China's rapid growth seemed ever more intractable, hope was in increasingly short supply, and the prospects for human and national flourishing were growing dim.[7] The contrast between the upbeat, utopian rhetoric of a science-obsessed leadership and the reality of life for scientists, engineers, and ordinary people on the ground was striking.

This volume presents the results of our discussions. Based on research conducted between 2006 and 2018, the chapters explore the makings, workings, and effects of various sciences and technologies.[8] Our focus is on an array of applied health and environmental knowledges and innovations being developed to solve some of the gravest problems of human and ecological health facing China today. The kinds of cutting-edge basic sciences that are being energetically promoted by the state and private entrepreneurs remain a subject for future anthropological research (but see Ong 2016, 197–222). The approach here is ethnographic in being based on fieldwork in China (as well as documentary, visual, and other modes of research), and in seeking to capture and reflect the actors' own points of view. Our informants include both experts (scientists, technicians, surgeons, therapists) and ordinary Chinese (pollution sufferers, patients, and other categories of citizens).

Governing China through Science: New Understandings of the 2000s and 2010s

Since the turn of the century, the anthropological study of China has been profoundly transformed by analyses of governance and governmentality. (Briefly, governance can be understood as efforts to shape conduct by agents within and beyond the state; governmentality is the modern regime of government that takes the optimization of the population as its primary end.) Focusing on logics, discourses, subjectivity, and other analytically key features of modern power, these studies have revealed how the market has joined the party-state to become arguably the central forces involved in governing Chinese society and creating human subjects (see for example, Li and Ong 2008). In this book, we make two major intellectual interventions. First, under the rubric "governing through science," we extend the governance/governmentality approach to the study of Chinese science and technology. Second, we deepen the analysis by adding the insights of science and technology studies. These analytic moves have important implications for both China studies and science studies. They greatly complicate the study of contemporary China by adding science to the cluster of governing agents, and placing the hugely complex but little-understood nexus of state-market-science/technology at the very center of the governance of social life. By focusing on an array of problems of government, we also push STS beyond its current preoccupations with such issues as actor network theory and ontology to consider problems of life-and-death importance in countries of the Global South (the erosion of human and environmental health, for example). Although this is not the place for a detailed discussion of the governance/governmentality and STS perspectives, a few basic ideas and orienting terms should help guide readers who may be unfamiliar with these bodies of thought, while situating our project in relation to a larger theoretical literature. These constructs were originally developed to understand

Western liberal societies but, with some modification, have proven fruitful in understanding China as well.

Science and Technology as Instruments of Governance

Understanding *governance* broadly as the "conduct of conduct," work on modern governance focuses on governmental projects—that is, more or less rationalized schemes undertaken by multiple authorities (state bureaucrats, professional experts, self-governing citizens), employing a variety of knowledge forms, that seek to influence conduct according to specific norms in order to achieve certain ends, with diverse and mostly unpredicted effects (foundational theoretical texts include Burchell, Gordon, and Miller 1991; Dean 1999; Foucault 1991; Rose 1999). We will encounter many such rationalized and consequential schemes below, from the Ministry of Health's program to build a community-level mental-health infrastructure (Ma, chapter 1), to a surgeon's game plan for making China a center of experimental stem-cell transplantation (Song, chapter 3), to an independent scientist's efforts to promote the black soldier fly as the solution to problems of urban waste management (Amy Zhang, chapter 7). The brief analytic sketch above provides a way to think about these projects: how they are structured, who counts as a "governor," what they do, and so on.

A modern governance perspective emphasizes the importance of rationalities of government, especially knowledge- or science-based ones. As the core logic in modern systems of governance, science shapes governing in countless ways. Science and its language of numbers often supply the rationale behind governmental projects and the authoritative norms those projects promote. Because of their status as authoritative knowledge producers, scientists, both human and natural, are often active participants in the political and policy process. In an authoritarian system such as China, where scientists are subject to party-state controls, they are expected to serve the party and government by lending their expertise to the making of official policies and plans (Cao and Suttmeier 2017). Indeed, as noted earlier, scientific policymaking is mandatory, making experts and expertise essential parts of the policy process. And there is more, for science is the ultimate arbiter of truth in modern societies; when science speaks in the name of nature, it depoliticizes objects of inquiry that may be profoundly political and thereby removes them from the field of contestation. These political capacities of science are especially pronounced in China, whose state has always treated S&T as tools of state power and whose ruling party has staked its legitimacy on its claim to be a scientific and technological modernizer capable of engineering the use of S&T to achieve national wealth, power, and global status. For China's party—and of course its people—the stakes in the making, claiming, and applying of science and technology could not be greater.

Science as Politics by Another Name: New Layers of Meaning

The interdisciplinary field of STS highlights the political nature of knowledge-making and the permeable line separating science from politics. In the early days of the field, these observations were captured in the pithy phrases of Donna Haraway and Bruno Latour, who famously declared: "Science is politics by other means" (Haraway 1984; Latour 1983, 1988). Over the years, STS scholars working in Western liberal societies have imbued the notion of science-as-politics with a multiplicity of meanings. Science-as-politics has come to refer to the contestation among ideas, for example, or the embedding of differences along lines of race/class/gender/sexuality in scientific thought. After thirty-some years in circulation, the notion has become something of a truism in the field.

In authoritarian China, the notion of science as politics takes on still more layers of meaning, for science and the party-state are intimately connected. As Cong Cao and Richard Suttmeier explain, in the West, a high degree of autonomy from political pressure is seen as necessary for the responsible exercise of scientific expertise expected by society. In China, by contrast, since the founding of the People's Republic in 1949, "professional autonomy has been circumscribed and viewed as antithetical to the political formula of the CCP [Chinese Communist Party]" (Cao and Suttmeier 2017, 1021). This additional meaning of science-as-politics—the subordination of science to governing authorities and their agendas—deserves our closest attention, especially because the relationship between science and the party-state is not stable or static, but rather always shifting. Indeed, in the Xi era, when S&T have been assigned vital roles in national rejuvenation, party control over the scientific and technical communities appears in some ways to be growing (Cao and Suttmeier 2017). The political urgency surrounding the promotion of S&T is rooted in the reality that the party's number one priority is remaining in power; its primary strategy for doing that is maintaining high economic growth rates; and the key to that, party leaders believe, is advanced S&T. Thus China's sciences and technologies serve a political master with an overriding interest in delivering the economic goods that will keep the people rich and content. Put another way, in the making of Chinese sciences and technologies, both the political and the economic demands of the party-state loom large. The subordination to the party-state is evident in many ways. Many if not most scientists and engineers work in state-run (and party-overseen) organizations, and the party-state possesses countless means, formal and informal, material and ideological, by which it can influence what counts as truth and how truth is made and promoted, even among those in private-sector positions (see, for example, Cao 2014; Hong and Zhao 2016; Tenzin 2017). How and under what circumstances these mechanisms operate are poorly understood, making in-depth ethnographic research on science-making vitally important.

In this volume, the micropolitics of science-making receives particular attention, as our contributors dissect the social dynamics by which their expert-informants gather data, fashion concepts, deploy measures, and promote their findings. Beyond the more quotidian discussions of data sources, quality, presentation, and the like, we show that in China scientists and engineers often find themselves negotiating the nuts and bolts of science-making and science-advancing with public officials. Environmental scientists are constrained to work out the parameters of their research with local cadres (see Lord, chapter 5); public health researchers must massage politically correct data supplied by their leaders into something resembling the truth (Mason, chapter 4); and scientists developing low-tech, traditional Chinese approaches to waste management must frame their projects as high-tech, modern, and commercially viable to make their ideas comprehensible in the scientistic and economistic discourse of the regime (Amy Zhang, chapter 7). Indeed, the research presented below suggests that negotiations with officials of the party-state may be simply a routine part of science- and technology-making in China. More broadly, our work shows that the relationships between the scientific community and agents of the party-state, far from simple subordination, are nuanced and negotiated in ways that need to be better understood.

Science Is Contextual: Politics and Economics in Command

For students of STS, science is no one thing; instead, it is humanly made in such a way that it bears the fingerprints of its makers and of the context in which it is made. And indeed, we will see in the chapters that follow that Chinese science is distinctly Chinese, bearing the imprint of unique historical and cultural forces at local and national levels and of the nation's place in global scientific and political-economic hierarchies.[9] Two key features of the wider context stand out in the chapters. The first is the prominence of market logics, which is a result of the decline in state funding for research and the state's push to commercialize academic knowledge, as well as the predominance of economic goals in Chinese politics generally. In one case, an environmental scientist called on to "innovate through commercialization" had to set aside his basic research to focus on developing marketable products and raising capital to scale up production (Amy

Zhang, chapter 7). In China, where state regulation of business is weak, the power of market forces can at times be virtually unchecked, putting great pressure on experts who are trying to fashion scientific and technological solutions to the nation's problems. In the most extreme cases presented in this book, environmental and public health researchers had to submit to market logics simply to survive. In one case, researchers were subject to the economic growth imperative of local-level leaders (Lord, chapter 5); in another, Chinese scientists had to subordinate their work to the profit imperative of foreign firms (Greenhalgh, chapter 6). In such cases, science itself could be said to be marketized.

The second is China's still very subordinate position in the global hierarchy of science. Though deeply rooted power imbalances constrain the development of Chinese science in a myriad ways, the chapters in this volume focus on how Chinese science is perceived by Chinese and foreign observers—almost invariably (though this has begun to change) as backward relative to that of the West—and how Chinese researchers and citizens attempt to right the global order of things. In several chapters, Western (as well as Japanese) sciences and technologies, considered superior by definition, are ardently embraced by Chinese experts and citizens searching for solutions to urgent social problems (Li Zhang, chapter 2; Mason, chapter 4; Greenhalgh, chapter 6; Kohrman, chapter 8). In some cases Western knowledges are praised even though they may not be well understood (Mason, chapter 4). The experts who are embracing and reworking foreign sciences are driven not only by a need for scientific solutions but also by a deep desire to catch up with the advanced nations and to show off the prowess of Chinese S&T so that Chinese experts will be accepted as equal members of the global scientific community. In yet another chapter, foreigners' criticisms of Chinese research and practice as ineffective, fraudulent, and even unethical provoke Chinese practitioners to defend their work by developing novel assessment tools and ethical formulations that fit the Chinese context (Song, chapter 3). In virtually every domain of science and technology examined, Chinese experts struggle mightily to be accepted as full members of their worldwide community of practitioners. Yet the going is tough, and they never quite succeed.

The Extraordinary Productivity of Science and Technology in China

If science carries the stamp of its context, that context also bears the stamp of science. Another fundamental tenet of STS is that modern sciences and technologies are highly consequential, or productive, shaping how life is lived and society is organized. In the language of the field, science and society (or science and politics) are *co-constituted*, produced in the same moment and in relation to one another (Jasanoff 2004). The co-production notion is especially illuminating in China, where science has the backing of a still-powerful party-state that, despite the rise of the internet and social media, largely dominates public discourse. As we will see, the result is that state-supported sciences and technologies have enormous force to shape not only the plans and policies of the party-state, but also the worldviews and personal identities of the people. This insight was a major contribution of the earliest work in the anthropology of Chinese S&T, which sought to put science and statistics on the intellectual map of the field. Korhman's (2003, 2005) genealogy of disability statistics illuminated how a massive 1987 survey, which for the first time enumerated the disabled population, served both to secure the biologized identities of disabled officials and to build a bio-bureaucracy with the legitimacy and ability to meet their needs. In my account of the historically first major instance of the newly mandated scientific policymaking (Greenhalgh 2003, 2008), I revealed that the inner party struggles over the one-child policy that erupted in 1979–1980 were at root contests over which science of population would shape party policy. Through its impact on the one-child policy—the policy's crisis rationale, tight targets, and blindness to gender and other social consequences—the winning science of population cybernetics profoundly reordered Chinese society and politics, creating jagged distortions in the social structure that planners are still struggling to correct.

The chapters in this volume make clear that, despite the major administrative and governmental shifts that have occurred in China since the early post-Mao era dissected in the above works, the impact of science is equally pronounced today. The effects are particularly visible in the field of mental health, which in recent years has been the target of sustained efforts at multiple levels to find scientific solutions to soaring rates of psychological distress and untreated mental disorder. At the central government level, the administrative creation of a huge new network of numbers has worked to construct communities as objects of governance, create populations of sufferers, distribute those populations in marked territories, and enable mental-health specialists to monitor and serve them in their communities (Ma, chapter 1). After this intervention, the practice of mental health in China will never be the same. The transformative nature of psychological science is also evident at the popular level, where therapists—the new experts in human emotions and relationships—are gaining the authority to define the good life and the good family for their middle-class clients (Li Zhang, chapter 2). These chapters reveal what is at stake in our study of S&T today. Taken together, the chapters in this book suggest too that in the China case, the co production idiom may be necessary but not sufficient to capture the mutual productivities of two domains of reality ("politics" and "science") that are not only deeply interdependent but also hierarchically ordered.

New knowledges and technologies not only create novel forms of sociality (populations of the mentally ill, notions of normative personhood), but also interact with existing social realities, altering them in the process. The chapters that follow highlight intersections with entrenched social divides (rural/urban, male/female, rich/poor), showing how new knowledges and gadgets may reproduce and even worsen old inequalities. Environmental research, which is supposed to ease the glaring ruralization of pollution, ends up reinforcing the ecological burden imposed on China's villages (Lord, chapter 5). Technologies of air pollution, which once gave women power over their smoking husbands, are now stirring up feminist anger as husbands light up inside air-purified homes (Kohrman, chapter 8). Costly new therapies for mental and physical health are privileging middle-class over working-class and foreign over Chinese patients, offering succor to some while denying it to those who may need it most (Li Zhang, chapter 2; Song, chapter 3). What these ethnographic cases suggest is that the much-celebrated modern S&T may be making an already unequal nation even more so. Science and technology are important parts of the story behind China's gaping socioeconomic divides.

Focusing on individual, public, and environmental health, this volume explores three sets of questions. First, how do these dreams and schemes for better health and lives through S&T circulate through Chinese society? Which dreams are still alive and which are dying? Second, how are the sciences of physical, mental, and environmental health made and made to fit to Chinese realities? How does the official elevation of modern S&T shape how science and technology are constructed and applied to resolve pressing problems of the day? Third and finally, are the party's and people's dreams of personal and national salvation coming true? What effects—intended and otherwise—are those practices having on China's politics, society, human and environmental health, and personhood in a time of rapid and profound societal transformations?

By placing the science question at the center of the study of contemporary Chinese society, this book aims to discover how different China might look when science and technology are given their due and in the process make science and society more central to the intellectual map of late twentieth- and early twenty-first-century China. By providing ethnographic insight into the making, workings, and effects of various sciences and technologies, we also aim to provide scholars in science and technology studies with up-close accounts and analytic frameworks for understanding Chinese S&T based on in-depth knowledge of the country's distinctive history, socioculture, and political economy. As a rising global superpower, one with multiplying connections to and effects on science communities the world over, China is a critical case for the field. A close study of China can also contribute to the development of a transnational field of science studies by illuminating both the workings of science in a non-Western and nondemo-cratic society and the connections among sciences in different parts of a rapidly changing world.

Are Modern Science and Technology Saving China? A Look Ahead

The chapters below are arranged in four pairs. Each pair supplies a different part of the answer to our overarching question of whether science and technology, as currently configured, are capable of "saving" China from the human and eco-logical fallout of four decades of growth at any cost. In the first pair of chapters, two scholars explore the great faith that government planners and ordinary people alike place in the promise of modern science to alleviate the many crises plagu-ing the country. Both chapters focus on the mounting crisis of mental distress. Zhiying Ma investigates a Ministry of Health program launched in 2004 to deliver basic mental-health care to communities nationwide. Focusing on the program designers and psychiatrists, she uncovers the supple if invisible role of numbers in the creation of a mental-health infrastructure. Yet the numbers deliver more than anyone expected. Ma shows how the planners' dream—of perfect, grid-like numerical governance capable of surveilling and serving all—not only becomes a nightmare of plan-targets and social control at the grassroots but also, by ex-cluding common disease categories, fails to meet the needs of large swaths of the population.

Shifting to the popular level, Li Zhang charts how, since around 2000, happi-ness and psychological well-being more generally have emerged as key compo-nents of "the good life" (*meihao shenghuo*) and how the Western sciences of psy-chology and psychotherapy are seen as the key instruments for achieving these desired states. But in this fuzzy area of human science, the meaning and efficacy of science are anything but clear cut. Zhang presents a series of ethnographic en-counters between counselors and patients, showing how therapists take advan-tage of the popular faith in science to frame their sometimes unproven approaches as scientific; how contests over efficacy are routine parts of the therapeutic en-counter; and how discouraged patients respond when the promise of happiness is not fulfilled. Can science help solve the growing problems of mental disease and distress in China today? These two chapters suggest that the hope remains very much alive, but whether science helps depends on who gets to define the meanings and uses of "science" and where one stands in relation to the science project. In these cases, the benefits of mental health science accrue not to the sufferers, but to the planners and counselors in charge of delivering the science to the people.

In the next pair of chapters, two contributors shed light on the struggles of Chinese scientists and clinicians to produce "objective," internationally credible scientific results in the challenging context of postsocialist China. In her study of a pioneering but controversial fetal stem-cell-transplant practice aimed at helping patients with ALS (Lou Gehrig's disease) and others facing rapid neuromuscular decline, Priscilla Song shows how the surgeon in charge sought to rebut Western accusations of lack of scientific rigor, poor ethics, and even quackery by developing alternative standards of proof, measures of efficacy, and ethical stances. Though he aspired to live up to ideal scientific and ethical principles, his efforts were undermined by the realities of a society in perpetual social and communicational flux.

Turning to public health science, Katherine Mason unearths the multiple forms of scientific truth created in a local Chinese Center for Disease Control and Prevention (CDC). She shows how a younger generation of scientists trained abroad, who initially hoped to contribute to the production of (what they considered) pure, untainted, internationally accepted scientific truth, discover to their frustration that they are trapped in a system dominated by their less educated elders in which the production of data is deeply shaped by social networks and clientelist political obligations.[10] Hoping to do "real science" that could save China both by preventing another severe acute respiratory syndrome (SARS) epidemic and by boosting the global reputation of the nation's public health community, they find themselves able to produce only "good-enough truths" that remain inferior or even false.[11] In both chapters, ambitious researchers seeking to do good science that solves problems and enhances China's reputation find themselves stymied by a social, political, and cultural context inimical to those lofty ends.

The third pair of chapters exposes some of the hidden dangers of China's highly commercialized sciences. These two chapters show how the widespread marketization of science—both the mandate to create economically useful science and the need to rely on non-state, including corporate, funding—leads to science that is fragmentary at best and practically ineffective or even harmful at worst. In her account of environmental science-making, Elizabeth Lord shows how the prioritization of economic over environmental goals that has long dominated the political process is reproduced in the science-making process. Constrained to fit environmental studies into a profit logic, researchers find themselves subject to rural cadres' demands to prioritize economic growth, prevented from gathering data from the most polluted villages, and dependent on political connections to do any research. The result is a distorted body of knowledge that excludes concerns with environmental justice, fails to address rural pollution control, and ends up reproducing the very gap between urban and rural that environmental science and policy are supposed to address.

In my analysis of the making of obesity science and interventions (Greenhalgh, chapter 6) I reveal how scientists' dependence on multinational soda companies for funding leads to the inadvertent adoption of scientific ideas and policies that may have protected soda profits but did little to arrest the rapid increase in obesity and related chronic diseases. The danger here is compounded by a lax environment around scientific ethics, in which the lead researchers are able to frame their embrace of corporate-funded projects as fully compliant with Chinese research ethics because the state's ethical bar is so low and its support for global capital is so strong. These chapters suggest that when science becomes subordinate to economic demands and the state underfunds science while placing few ethical or other limits on corporate intervention, real solutions to urgent problems such as ecological degradation, rural decay, and soaring rates of chronic disease become nearly impossible. In any case, the sort of marketized science that is widespread in China today contains few answers to the vexing problems that trouble the nation.

In the fourth and final pair of chapters, two scholars trace what happens when the state's promises of a better life through modern S&T palpably fail to deliver, and public faith in state solutions ebbs. One response has been the creation and often difficult promotion of indigenous technological solutions. China is drowning in its urban waste and much of it is organic matter. In her chapter on waste management, Amy Zhang charts the public's (and experts') growing distrust of the imported large-scale infrastructural solutions favored by the state, all of which have proven unviable or even toxic. In the 2010s some researchers have been reaching back to an older tradition of Chinese entomological science that saw insects not as public health threats to be eradicated, but as potential resources for human betterment. Zhang's chapter tells the story of an independent scientist who is developing a low-tech solution that relies on adult flies to devour organic waste, while marketing the larvae as protein-rich animal feed. Early evidence for its effectiveness is promising, and some state support can be found in official discourses on "indigenous innovation," which are part of China's drive for national autonomy. Yet whether the insect solution can be scaled up and succeed in the political context of Xi's China remains highly uncertain, for state discourses on science include the imperative of commercialization, which forces scientists to become business-minded entrepreneurs. In principle, indigenous innovation sounds promising, but in the political context of contemporary China it is likely to offer a partial way out at the very best.

A second response is closer to despair. The specter of slow death by airborne particulate matter—the dreaded PM2.5, whose levels have far exceeded safe levels for years—has led to an air filtration craze among urban middle-class households desperate to filter out the pollution that is quietly eroding their health. In his rumination on the historical and emotional links between two technologies of filtration—the filter-tipped cigarette introduced in the 1980s and the air purifier of today—Matthew Kohrman excavates the gendered politics of action and affect these technologies have spurred. He shows how the urban home, once a battleground over smoking, has become a war zone over air purification, as men embrace the purifying machines as promising high-tech solutions, while women despise and distrust them, expressing a tangle of fear, endangerment, desperation, and hopelessness over being held captive in their homes like "caged birds." Kohrman's chapter is a sobering reminder of the social and emotional costs imposed on China's people by the failure of the state's vaunted "modern S&T" to alleviate the severe environmental pollution that its policies on economic growth created.

In an afterword, Mei Zhan reflects on the book as a whole, highlighting the specificities of the sciences and technologies in China and the ethnographic and conceptual contributions China anthropology can make to transforming STS into a more truly global field.

Through deep dives into the micropolitics of knowledge and innovation, these chapters expose to daylight a yawning gap between the promises delivered by the party and the frustrations of ordinary people trying to live a good—or even just decent—life in China today. Although faith in science and technology has remained strong in most of the communities studied, expert and lay alike, in case after case we found that the promises attached to them have not been realized: Policies were ineffective or even harmful; programs furthered state control instead of popular health; treatments were rife with ethical and efficacy problems; initiatives reproduced existing inequalities; and the promised good life seemed forever postponed. This is not to say that none of the scientific and technological innovations developed to address China's problems has worked; certainly, many have, even if in unexpected ways. Still, in all the cases we subjected to an anthropological gaze, the gains were invariably shadowed by losses, the truths ruptured by paradoxes. Utopian dreams too often were followed by dystopian realities. Our analysis of the wider contexts shows that the failures of science to fix the targeted problems can be traced to the imprinting of party histories and politics, profit motives, and existing social inequalities on the science that was made. Science is, in short, too subservient to overarching economic and political agendas of the party that conflicted with the goals set for science. And that subservience, captured in the analytic of the state-market-science/technology nexus, appears to be intensifying under the ambitious, authoritarian leadership of Xi Jinping.

Our work on the ground in the People's Republic raises critical questions for future research. If Chinese science and technology are, in the end, mostly by and for the party-state and its agents, under what conditions can they also improve the lives of China's people? Which people in which places are most likely to benefit? Short of a drastic change in leadership, what changes in political-economic arrangements or sociocultural norms might precipitate a shift toward greater political independence for China's researchers and technicians? Can the impetus for change come from outside, or must it originate within China itself? To what extent do our conclusions, which are based on our study of practical sciences and technologies, apply to the more cutting-edge fields of S&T energetically supported by the state? These are just a few of the questions that we hope might provoke other researchers to make the sciences and technologies more central parts of their study of contemporary China.

NOTES

The author thanks Arthur Kleinman, Elizabeth Lord, Amy Zhang, Li Zhang, and two reviewers for the press for their insightful comments on earlier drafts of this introduction. The author is grateful to Wei Hong for many illuminating discussions of STS in China, and to Victor Seow for sharing his perspectives on the work of historians of science and technology in late twentieth-century China. His thoughts are reflected in note 4.

1. Xi Jinping holds an undergraduate degree in chemical engineering but advanced degrees in law and politics. Li Keqiang, Premier of the State Council and head of government, has a graduate degree in economics (Li 2016).

2. In March 2018, the Thirteenth National People's Congress amended the constitution to eliminate term limits on Xi's post as president, opening the way to his remaining in power for a great many years.

3. Some practices deviate so significantly from conventional notions of scientific activity that some observers may deem them unscientific or even anti-scientific. We avoid such language here. Rather than imposing an outsider's view of what is scientific and what is not, we examine what trained scientists in China present as "science," delving into how it is crafted and what work it performs.

4. Science and technology are growing areas of interest among historians of modern China, including those working in the early and mid-twentieth century (see, for example, Rogaski 2004; Lam 2011). In the last fifteen years, a few historians have begun to take up the Maoist era (1949–1976) as history. Their central concern has been to write against the narrative that "all Maoist science was bad science" (see especially Schmalzer 2008; Schmalzer 2016; Wei and Brock 2013). By showing ways in which certain scientific practices and projects worked, they have been considering whether there is something distinct that can be called "socialist science."

5. In addition to the work of STS-influenced anthropologists discussed in this book, in-depth scholarship on China within STS includes writings on regulatory and ethical governance of stem-cell research (Sleeboom-Faulkner 2015; Sleeboom-Faulkner and Sui 2015; Zhang 2012). Much of this STS work concerns China's life sciences. Unfortunately, institutional and other constraints have greatly slowed the development of STS scholarship in China itself. The assessment of Liu Bing (2011) remains relevant today.

6. This is not intended as a comprehensive list of anthropological writings on science and technology in China. The publications just cited include the major book-length studies and, for scholars whose work on science (and technology) has appeared so far only in article or chapter form, a key article.

7. Such a bleak outlook may have dominated our discussions because the papers dealt with the difficulties science and technology have encountered in solving social and environmental problems. China's achievements in high-tech engineering (visible in major infrastructural developments, for example) and the frontier sciences have spurred great national pride.

8. Scholars in STS often use the term "technoscience" to signal that science and technology are not readily differentiated (with one engaging in basic research, the other in applied practice) and should be understood not as separate fields but as co-constructed and hybrid forms of knowledge and practice (Latour 1987). This term has not yet caught on in the study of Chinese knowledge practices. Though the term may be suitable for the analysis of China's cutting-edge sciences and innovations, in this volume we deal mostly with simpler, applied sciences and technologies in which the connections between knowledge and application are a small part of our accounts. For that reason we follow conventional practice in China studies and refer to "science and technology."

As this book goes to press in early 2019, China's great leap into artificial intelligence, which took off in the mid-2010s, appears to be rapidly accelerating the societal relevance of technoscientific logics and practices. Today automated machine processes using algorithms created through machine learning from massive amounts of personal data scraped from networked smartphones, surveillance cameras, and other devices are shaping individual behavior in ever more domains of Chinese life (Lee 2018). I discuss these developments elsewhere.

9. China of course is not alone in this; since all science formations reflect their wider context, all national science systems can be described as unique to the host nations. Whether there is a universal science with shared values (such as truth-seeking and freedom of inquiry) is a different question, one that has garnered considerable interest in the China context, where such values are not much in evidence (e.g., Cao 2013, 155).

10. An especially illuminating case study of the multiple levels of hierarchy and subordination in the field of geosciences is Hong (2008).

11. The 2002–2003 SARS outbreak in southern China led to eight thousand cases and over seven hundred deaths worldwide, with the majority in China. The hostility met by the government's initially poor handling of the epidemic led to major changes in how China handles infectious-disease threats to public health (Mason 2016 and chapter 4 of this volume).

References

Andreas, Joel. 2009. *Rise of the Red Engineers: The Cultural Revolution and the Origins of China's New Class.* Stanford, CA: Stanford University Press.

Baum, Richard. 1980. *China's Four Modernizations: The New Technological Revolution.* Boulder, CO: Westview.

Berlant, Lauren. 2011. *Cruel Optimism.* Durham, NC: Duke University Press.

Bunkenborg, Mikkel. 2014. "Subhealth: Questioning the Quality of Bodies in Contemporary China." *Medical Anthropology: Cross-Cultural Studies in Health and Illness* 33(2): 128–143.

Burchell, Graham, Colin Gordon, and Peter Miller, eds. 1991. *The Foucault Effect: Studies in Governmentality*. Chicago: University of Chicago Press.

Cao, Cong. 2004. *China's Scientific Elite*. London: Routledge Curzon.

Cao, Cong. 2014. "The Universal Values of Science and China's Nobel Prize Pursuit." *Minerva* 52(2): 141–160.

Cao, Cong, Ning Li, Xia Li, and Liu Li. 2013. "Reforming China's S&T System." *Science* 341(6145): 460–462.

Cao, Cong, and Richard P. Suttmeier. 2017. "Challenges of S&T System Reform in China." *Science* 355(6329): 1019–1021.

Carlson, Benjamin. 2015. "The World According to Xi Jinping." *Atlantic*, September 21. https://www.theatlantic.com/international/archive/2015/09/xi-jinping-china -book-chinese-dream/406387/.

Chen, Ling, and Barry Naughton. 2016. "An Institutionalized Policy-Making Mechanism: China's Return to Techno-Industrial Policy." *Research Policy* 45(10): 2138–2152.

Chen, Nancy N. 2003. *Breathing Spaces: Qigong, Psychiatry, and Healing in China*. New York: Columbia University Press.

Chen, Nancy N. 2014. "Between Abundance and Insecurity: Securing Food and Medicine in an Age of Chinese Biotechnology." In *Bioinsecurity and Vulnerability*, edited by Nancy N. Chen and Lesley A. Sharp, 87–102. Santa Fe, NM: SAR Press.

Chinese Academy of Sciences (CAS). 2016. "Xi Sets Targets for China's Science, Technology Mastery," May 31, http://english.cas.cn/newsroom/news/201605/t20160531_163783.shtml.

Cyranoski, David. 2016. "What China's Latest 5-Year Plan Means for Science." *Nature*, March 18. http://www.nature.com/news/what-china-s-latest-five-year-plan -means-for-science-1.19590.

Dean, Mitchell. 1999. *Governmentality: Power and Rule in Modern Society*. London: Sage.

Economy, Elizabeth C. 2018. *The Third Revolution: Xi Jinping and the New Chinese State*. New York: Oxford University Press.

Elman, Benjamin A. 2005. *On Their Own Terms: Science in China, 1550–1900*. Cambridge, MA: Harvard University Press.

Farquhar, Judith. 1996. *Knowing Practice: The Clinical Encounter of Chinese Medicine*. Boulder, CO: Westview Press.

Farquhar, Judith. 2002. *Appetites: Food and Sex in Post-Socialist China*. Durham, NC: Duke University Press.

Fewsmith, Joseph. 2008. *China since Tiananmen: From Deng Xiaoping to Hu Jintao*, 2nd ed. Cambridge, UK: Cambridge University Press.

Five-Year Plan. 2016. Thirteenth Five-Year Plan for Economic and Social Development of the People's Republic of China. Beijing: Central Compilation and Translation Press. http://en.ndrc.gov.cn/newsrelease/201612/P020161207645765233498.pdf.

Foucault, Michel. 1999. "Governmentality," in Graham Burchell, Colin Gordon, and Peter Miller, eds., *The Foucault Effect: Studies in Governmentality* (Chicago: University of Chicago Press), 87–104.

Fuller, Douglas B. 2016. *Paper Tigers, Hidden Dragons: Firms and the Political Economy of China's Technological Development*. Oxford, UK: Oxford University Press.

"The Future of Chinese Research." 2016. *Nature*, June 22. http://www.nature.com/news /the-future-of-chinese-research-1.20123.

Goldman, Merle, and Roderick MacFarquhar, eds. 1999. *The Paradox of China's Post-Mao Reforms*. Cambridge, MA: Harvard University Press.

Gottschang, Suzanne. 2018. *Formulas for Motherhood in a Chinese Hospital*. Ann Arbor: University of Michigan Press.

Greenhalgh, Susan. 2003. "Science, Modernity, and the Making of China's One-Child Policy." *Population and Development Review* 29(2): 163–196.

Greenhalgh, Susan. 2008. *Just One Child: Science and Policy in Deng's China.* Berkeley: University of California Press.

Greenhalgh, Susan, and Edwin A. Winckler. 2005. *Governing China's Population: From Leninist to Neoliberal Biopolitics.* Stanford, CA: Stanford University Press.

Haraway, Donna. 1984. "Science Is Politics by Other Means." *In These Times*, October 10–16, 13, 15.

Hathaway, Michael J. 2013. *Environmental Winds: Making the Global in Southwest China.* Berkeley: University of California Press.

Hong, Wei. 2008. "Domination in a Scientific Field: Capital Struggle in a Chinese Isotope Lab." *Social Studies of Science* 38(4): 543–570.

Hong, Wei, and Yandong Zhao. 2016. "How Social Networks Affect Scientific Performance: Evidence from a National Survey of Chinese Scientists." *Science, Technology, and Human Values* 41(2): 243–273.

Hua, Shiping. 1995. *Science and Humanism: Two Cultures in Post-Mao China (1978–1989).* Albany: State University of New York Press.

Hyde, Sandra Teresa. 2007. *Eating Spring Rice: The Cultural Politics of AIDS in Southwest China.* Berkeley: University of California Press.

Jasanoff, Sheila, ed. 2004. *States of Knowledge: The Co-Production of Science and Social Order.* London: Routledge.

Kohrman, Matthew. 2003. "Why Am I Not Disabled? Making State Subjects, Making Statistics in Post-Mao China." *Medical Anthropology Quarterly* 17(1): 5–24.

Kohrman, Matthew. 2005. *Bodies of Difference: Experiences of Disability and Institutional Advocacy in the Making of Modern China.* Berkeley: University of California Press.

Kwok, D. W. 1965. *Scientism in Chinese Thought, 1900–1950.* New Haven, CT: Yale University Press.

Lam, Tong. 2011. *A Passion for Facts: Social Surveys and the Construction of the Chinese Nation-State, 1900–1949.* Berkeley: University of California Press.

Lamoreaux, Janelle. 2016. "What If the Environment Is a Person? Lineages of Epigenetic Science in a Toxic China." *Cultural Anthropology* 31(2): 188–214.

Latour, Bruno. 1983. "Give Me a Laboratory and I Will Raise the World." In *Science Observed: Perspectives on the Social Study of Science*, edited by Karin D. Knorr-Cetina and Michael Mulkay, 141–170. Los Angeles, CA: Sage.

Latour, Bruno. 1987. *Science in Action.* Cambridge, MA: Harvard University Press

Latour, Bruno. 1988. *The Pasteurization of France.* Cambridge, MA: Harvard University Press.

Lee, Kai-Fu. 2018. *AI Superpowers: China, Silicon Valley, and the New World Order.* Boston: Houghton Mifflin Harcourt.

Lee, Sing. 2011. "Depression: Coming of Age in China." In *Deep China: The Moral Life of the Person*, by Arthur Kleinman, Yunxiang Yan, Jing Jun, Sing Lee, Everett Zhang, Pan Tianshu, Wu Fei, Jinhua Guo, 177–212. Berkeley: University of California Press.

Lewin, Arie Y., Martin Kenney, and Johann Peter Murmann, eds. 2016. *China's Innovation Challenge: Overcoming the Middle-Income Trap.* Cambridge, UK: Cambridge University Press.

Li, Cheng. 2001. *China's Leaders: The New Generation.* Lanham, MD: Rowman and Littlefield.

Li, Cheng. 2016. *Chinese Politics in the Xi Jinping Era: Reassessing Collective Leadership.* Washington, DC: Brookings Institution Press.

Li, Cheng. 2017. *The Power of Ideas: The Rising Influence of Thinkers and Think Tanks in China*. Singapore: World Scientific.

Liu, Bing. 2011. "Advantages and Disadvantages: Some Reflections on Philosophy and STS Studies in Mainland China." *East Asian Science, Technology and Society* 5(1): 67–72.

Lora-Wainwright, Anna. 2013. "The Inadequate Life: Rural Industrial Pollution and Lay Epidemiology in China." *China Quarterly* 214: 302–320.

Mason, Katherine A. 2016. *Infectious Change: Reinventing Chinese Public Health after an Epidemic*. Stanford, CA: Stanford University Press.

Naughton, Barry. 2018. "China's Great Gamble." Lecture delivered at the Fairbank Center for Chinese Studies, Harvard University, China Economy Lecture Series, March 29.

Naughton, Barry. 2019 "Grand Steerage: The Temptation of the Plan." In *China's Paths to the Future*, eds. Thomas Fingar and Jean Oi. Stanford, CA: Stanford University Press.

Nie, Jing-Bao, Yu Cheng, Xiang Zou, Ni Gong, Joseph D. Tucker, Bonnie Wong, and Arthur Kleinman. 2018. "The Vicious Circle of Patient-Physician Mistrust in China: Health Professionals' Perspectives, Institutional Conflicts of Interest, and Building Trust through Medical Professionalism." *Developing World Bioethics* 18(1): 26–36.

Ong, Aihwa. 2016. *Fungible Life: Experiment in the Asian City of Life*. Durham, NC: Duke University Press.

Palmer, David A. 2007. *Qigong Fever: Body, Science, and Utopia in China*. New York: Columbia University Press.

Rogaski, Ruth. 2004. *Hygienic Modernity: Meanings of Health and Disease in Treaty-Port China*. Berkeley: University of California Press.

Rose, Nikolas S. 1999. *Powers of Freedom: Reframing Political Thought*. Cambridge: Cambridge University Press.

Saich, Tony. 1989. *China's Science Policy in the 80s*. Manchester, UK: Manchester University Press.

Schmalzer, Sigrid. 2008. *The People's Peking Man: Popular Science and Human Identity in Twentieth-Century China*. Chicago: University of Chicago Press.

Schmalzer, Sigrid. 2016. *Red Revolution, Green Revolution: Scientific Farming in Socialist China*. Chicago: University of Chicago Press.

Shen, Xiaobai, and Robin Williams. 2005. "A Critique of China's Utilitarian View of Science and Technology." *Science, Technology & Society: An International Journal Devoted to the Developing World* 10(2): 197–223.

Sigurdson, Jon. 2006. *Technological Superpower China*. Cheltenham, UK: Edward Elgar.

Simon, Denis Fred, and Cong Cao. 2009. *China's Emerging Technological Edge: Assessing the Role of High-End Talent*. Cambridge, UK: Cambridge University Press.

Simon, Denis Fred, and Merle Goldman, eds. 1989. *Science and Technology in Post-Mao China*. Cambridge, MA: Harvard University Asia Center.

Sleeboom-Faulkner, Margaret. 2007. *The Chinese Academy of Sciences (CASS): Shaping the Reforms, Academia, and China (1977–2003)*. Leiden: Brill.

Sleeboom-Faulkner, Margaret, ed. 2015 *Stem Cell Research in Asia: Looking Beyond Regulatory Exteriors*. London: Routledge.

Sleeboom-Faulkner, Margaret, and Suli Sui. 2015. "Governance of Stem Cell Research and Its Clinical Translation in China: An Example of Profit-Oriented Bionetworking." *East Asian Science, Technology, and Society: An International Journal* 9(4): 397–412.

Song, Priscilla. 2017. *Biomedical Odysseys: Fetal Cell Experiments from Cyberspace to China*. Princeton, NJ: Princeton University Press.

Suttmeier, Richard P. 1980. *Science, Technology, and China's Drive for Modernization*. Stanford, CA: Hoover Institution Press.

Suttmeier, Richard P. 1989. "Conclusion: Science, Technology, and China's Political Future—A Framework for Analysis." In *Science and Technology in Post-Mao China*, edited by Denis Fred Simon and Merle Goldman, 375–396. Cambridge, MA: Harvard University Asia Center.

Suttmeier, Richard P., and Xiangkui Yao. 2004. "China's Post-WTO Technology Policy: Standards, Software, and the Changing Nature of Techno-Nationalism." *NBR* Special Report no. 7. Seattle, WA: National Bureau of Asian Research.

Tenzin, Jinba. 2017. "The Ecology of Chinese Academia." *China Quarterly* 231: 775–790.

Tilt, Bryan. 2014. *Dams and Development in China: The Moral Economy of Water and Power*. New York: Columbia University Press.

Wahlberg, Ayo. 2018. *Good Quality: The Routinization of Sperm Banking in China*. Berkeley: University of California Press.

Wang, Weiguang, ed. 2009. *Kexue fazhan guan gailun* [*Outline of the Scientific Concept of Development*]. Beijing: People's University of China Press.

Wei, Chunjuan Nancy, and Darryl E. Brock, ed. 2013. *Mr. Science and Chairman Mao's Cultural Revolution: Science and Technology in Modern China*. Lanham, MD: Lexington Books.

Xinhua. 2017. "Xi Sets Targets for China's Science." *Technology Mastery*, May 30. http://news.xinhuanet.com/english/2016-05/30/c_135399691.htm.

Yang, Jie. 2015. *Unknotting the Heart: Unemployment and Therapeutic Governance in China*. Ithaca, NY: Cornell University Press.

Yip, George S., and Bruce McKern. 2016. *China's Next Strategic Advantage: From Imitation to Innovation*. Cambridge, MA: MIT Press.

Yu, Xie, Chunni Zhang, and Qing Lai. 2014. "China's Rise as a Major Contributor to Science and Technology." *PNAS* (*Proceedings of the National Academy of Sciences*) 111(26): 9437–9442.

Zhan, Mei. 2009. *Other-Worldly: Making Chinese Medicine through Transnational Frames*. Durham, NC: Duke University Press.

Zhang, Everett Yuehong. 2015. *The Impotence Epidemic: Men's Medicine and Sexual Desire in Contemporary China*. Durham, NC: Duke University Press.

Zhang, Everett, Arthur Kleinman, and Tu Weiming, eds. 2011. *Governance of Life in Chinese Moral Experience: The Search for an Adequate Life*. London, UK: Routledge.

Zhang, Joy Y. 2012. *The Cosmopolitanization of Science: Stem Cell Governance in China*. New York: Palgrave Macmillan.

Zhang, Li. 2014. "Bentuhua: Culturing Psychotherapy in Postsocialist China." *Culture, Medicine, and Psychiatry* 38(2): 283–305.

Zhou, Yu, William Lazonick, and Yifei Sun, eds. 2016. *China as an Innovation Nation*. Oxford, UK: Oxford University Press.

NUMBERS AND THE ASSEMBLING OF A COMMUNITY MENTAL HEALTH INFRASTRUCTURE IN POSTSOCIALIST CHINA

Zhiying Ma

On October 9, 2013, a day before that year's World Mental Health Day, a news article brought the Chinese psychiatric profession to the center of public attention. Published in *Southern Metropolis Daily*, the article was entitled "'Apportioning' [*tanpai*] Quotas [*zhibiao*] of Mental Illnesses." It reported that in Zhengzhou, the capital of Henan Province, community mental health practitioners (CMHPs)[1] had received a curious task: finding two persons with serious mental illnesses—schizophrenia, schizoaffective disorder, bipolar disorder, paranoid disorder, mental retardation with psychosis, or epilepsy with psychosis—per thousand residents. Because these CMHPs belonged to administrative communities, or *shequ*, of different population sizes, the quotas they were assigned were proportionately different. For example, in a community with 35,398 residents, seventy-one patients with serious mental illnesses had to be discovered (Wang and Jin 2013).

According to the article, the CMHPs were required to solicit clues to suspected patients from local bureaucrats and other community residents. They were then asked to visit the suspected patients' homes in order to confirm their conditions and to register their information in a database. Several CMHPs complained that their home visits were often unwelcome because of families' concern with privacy, and that on occasion doors were even slammed in their faces. Moreover, despite their best efforts, it just seemed impossible to find so many patients and meet the quotas. In contrast to the quota of seventy-one patients, the CMHP in the aforementioned community could only find twelve patients. "I can't possibly register those who are not mentally ill as patients, right?" asked another CMHP rhetorically. However, in order to avoid potential institutional punishment, some CMHPs did choose to make up the numbers by counting as seriously mentally ill people who were just a bit "off."

The article then pointed out that this "quota apportioning" was a nation-wide phenomenon produced by the 686 Program (Wang and Jin 2013). The program was established by China's Ministry of Health in 2004, and it is officially called the "Program for Managing and Treating Serious Mental Illnesses, Run by Local [Governments] and Subsidized by the Central [Government]." The short form came from the funding, RMB 6.86 million (approximately USD 1 million), which the program received from the central government in its first year (PKU6 2012). The program's main goals, as stated by its leaders, are "to establish an effective mechanism to comprehensively prevent and control the violent behavior of patients with serious mental illnesses; to enhance the treatment rate and reduce the violence rate; to disseminate knowledge of mental illness prevention and treatment; and to spread the knowledge of systematic treatment of serious mental illnesses" (Ma et al. 2011). After patients are discovered, the plan is for CMHPs to visit them regularly, provide them with free or low-cost medications, and check on them lest they harm themselves or others. Simply put, the 686 Program seeks to build a mental health infrastructure that extends beyond the psychiatric hospital and reaches the public.

Regardless of the program's goals, the public was scandalized by its "quota apportioning" practice, especially the potential consequence of "making up patients" and subjecting them to unpleasant treatment. In response, Dr. Yu Xin, a leading expert involved in designing and overseeing the 686 Program, explained to journalists that the target detection rate (*jianchu lü*) was calculated based on the prevalence rate of serious mental illness discovered in epidemiological surveys.[2] A spokesperson from the Ministry of Health claimed that the rate was merely a guideline for provinces and cities; it was not—and should not be—imposed on health workers at the community level. Echoing these statements, administrators from the Zhengzhou Bureau of Health emphasized the necessity of the 686 Program for patients and the general public (Han and Ren 2013). These explanations did not satisfy critics like Huang Xuetao, a famous human rights advocate for psychiatric patients. She told reporters that practices of community mental health rendered medicine an administrative matter, deviating from medical ethics and scientific methods of epidemiology. By discovering patients and putting them under surveillance, she argued, these practices made mental health services measures of social control (Huang 2013).

Numbers, Communities, and Dreams of Governance in Postsocialist China

As we can see in the Zhengzhou incident, numbers such as the patient detection rate have played an important and controversial role in the design, implementation, and public perceptions of the 686 Program. While the program's designers and leaders insisted on the scientific nature of the target detection rate, as well as its legitimacy and "soft" quality in practice, grassroots practitioners saw the target as removed from local reality and imposing much bureaucratic pressure on them. Shored up by numbers, the program was seen by its designers and leaders as a good for mentally ill patients and the general public, but it was deemed by critics to be a tool of state control. Why, then, does the construction of a community mental health infrastructure hinge on the assignment, collection, and monitoring of numbers? How does the same set of numbers elicit different feelings and assume different modes of operations? Why is community mental health as a form of numerical governance a dream for some and a nightmare for others? These are the questions that this chapter seeks to answer.

Numbers have become increasingly important for governance throughout the contemporary world. For example, scholars have noticed a rising "audit culture" in neoliberal societies (Strathern 2000), a dominant "indicator culture" in international human rights monitoring (Merry 2016), and an overflow of "metrics work" in global health (Adams 2016). In these forms of numerical governance, numbers project an "aura of objective truth and scientific authority" (Merry 2016, 1), free from personal, political, and moral biases; they provide seemingly universal standards, rendering different social worlds comparable; and their apparent transparency summons bureaucratic accountability even in the absence of direct state interventions or hard international laws. In postsocialist China, the scientific quality of numbers has given them a special appeal. After all, science appears to government leaders, experts, and the populace alike as an antidote to the tendency of over-politicization during the Maoist era. It provides a global standard to diagnose what is wrong with China, as well as to discern how the nation should catch up with the world and achieve modernity (Wang 2008). A key object of governance that numbers have helped to constitute in postsocialist China is the population, a biological entity that needs to be managed by science (Greenhalgh and Winckler 2005), a "collective form of grouping that renders focused research, measurement, and intervention possible" (Cho 2013, 70). Through categorizing, counting, estimating, and projecting, numbers have produced entities ranging from general population overgrowth (Greenhalgh 2008) to specific groups such as the disabled (Kohrman 2003), the poor (Cho 2010), and the infectious (Mason 2016). Taking cues from these studies, this chapter will show how globally validated epidemiological estimates have constituted a population of seriously mentally ill patients in China, the target population of the 686 Program.

Note, however, that numerical governance in China is not necessarily new or simply global. With a long genealogy in Chinese statecraft, the same set of numbers can often operate in different modalities, commensurate various desires, and produce diverse effects or affects (Kipnis 2008).[3] For instance, Matthew Kohrman has shown that the codification and quantification of disability allowed Chinese officials to frame the country as simultaneously "backward" and "developing" (Kohrman 2003). Susan Greenhalgh has discovered that the one-child policy combined Western science of population growth modeling with socialist target setting and party-led mobilization to produce a "target obsession and numbers mania," which treated population as numbers and nothing else (Greenhalgh 2005a). Similar dynamics can be found in the 686 Program. As I will show, the categorization and enumeration of seriously mentally ill patients has allowed psychiatrists and policymakers to commensurate their desires to serve and surveil this population. The dream-like quality of globally circulating numbers has thus helped fashion a dream for a state that is simultaneously caring, secure, and stable. However, as seen in the Zhengzhou incident, when epidemiological estimates operate as target detection rates and program evaluation standards, grassroots practitioners, clients, and critics are reminded of the nightmare of the socialist planned economy. I will suggest that this unintended consequence comes from people's historical memory of numerical governance, as well as the existing bureaucratic pathways that guide numbers' traveling.

In adding to the literature on numerical governance, I will show that numbers may help constitute not only populations but also communities as objects of governance. Since the demise of the socialist work units in the 1990s, the Chinese state has sought to reorganize the social along a new axis called "community" (Bray 2006; Tomba 2014). While some scholars have focused on "community" as governing through individuals, families, and private property management agencies (Zhang 2012), others have emphasized the extension of state institutions in community construction (Heberer and Göbel 2011; Read 2012). Here, community mental health seeks to extend institutional psychiatry and integrate mental health work into state-sponsored public health endeavors. Following Nikolas Rose's idea that community is both "the territory of government" and "a means of government" (Rose 1996, 335), I will analyze the ways in which community is brought to bear on mental health and numbers help construct community. First, numbers can distribute the seriously mentally ill population in bounded territories. The specific geographical imagination of patient distribution is shaped by both the universal, abstract vision of epidemiological estimates and the particular considerations in target setting. Second, by counting out patients estimated to exist and recording their quantitative information, numbers enable profession-als to serve and surveil them in situ. Third, through target setting, patient enu-meration, and quantitative program evaluation, a team of CMHPs with common protocols and different areas of operations is built. Although targets are envisioned by their designers as no more than soft guidelines, once promoted, they obtain an inertia to travel along existing bureaucratic pathways, press on grassroots prac-titioners, and roll out the community mental health infrastructure down to the lowest administrative level.[4]

By exploring how numbers help construct the community mental health in-frastructure, this chapter also speaks to discussions about emerging health and therapeutic governance in China, especially evaluations of the Chinese state's re-newed promise to care. In her study on state-sponsored psychotherapeutic ser-vices for unemployed workers, Jie Yang (2015) uses the term "kindly power" (xiv) to debunk the "trick of kindness that is intended to disguise the unkindness of the state" (25), including past destructions caused by the state-led market reform and current social control. Similarly, Katherine Mason has revealed a "bifurca-tion of service and governance" in public health programs—that is, a separation "between the group being served and the group being governed" (Mason 2016, 20). On the other hand, Li Zhang (2017) has discovered that therapeutic governing can "simultaneously produce disciplining and nurturing, repressive and unfet-tering effects in everyday life" (6). She thus suggests that instead of assuming a cynical view, we pay attention to the visions driving local authorities to pursue therapeutic governance, as well as its diverse effects (9–10). In the case of com-munity mental health, I will argue that numbers provide a medium through which different dreams of governance—service *and* surveillance, humanitarian care *and* security management—are framed, justified, commensurated, and operational-ized.[5] Therefore, at least for program designers, psychiatric power exerted through community mental health is genuinely kind. Yet by tracing numbers' circulation, I will also show how service may be unhinged from surveillance, and why com-munity mental health governance may be perceived as unkind. Not only can pro-gram targets, transformed from epidemiological estimates, elicit fear, impose bureaucratic pressure, and incite tactical responses of data fabrication, but they may also outrun and limit services that CMHPs provide. Communities that num-bers carve out exclude certain groups in need of care, and this exclusion in turn disrupts the geographical imagination entailed in the distribution and assignment of numbers.

This chapter is structured as a genealogy of numbers (Merry 2016) in the 686 Program. In what follows, I trace what conditions and visions undergird the emer-gence of community mental health; how globally validated epidemiological esti-mates constitute the client population of the 686 Program and commensurate the program's different visions; how national and local interests translate these estimates into program targets and evaluation standards; how the numbers' circulation in existing bureaucratic pathways can generate controversies of "quota apportioning"; and, finally, in its everyday operations, what the numerically guided community mental health infrastructure might include or exclude, when it might work or break down. Data for this chapter draw on my ongoing ethnographic research on community mental health in China. They include interviews with the 686 Program's designers, leading experts, and local officials; a review of policy documents, news reports, and academic publications related to community mental health; and my fieldwork with CMHPs in the city of Nanhua, a southern Chinese metropolis with a resident population of well over 10 million, during 2013–2014.[6]

Service, Surveillance, and the Emergence of the 686 Program

I begin by tracing the emergence of the 686 Program and community mental health in China. During the Maoist era, China's health sector prided itself on an effective three-tier referral system and a large cadre of barefoot doctors, which together provided urban and rural residents with accessible and inexpensive primary care. In 1958, the Ministry of Health organized the First National Conference of Psychiatric Specialists in Nanjing, advocating for active prevention and treatment of mental illness in the form of localized outpatient care (rather than concentrated inpatient care) (Pearson 1995). Although psychiatry as a profession soon lost the favor of the Maoist state, psychiatrists kept experimenting with local models of community mental health service delivery up until the 1990s. However, thanks to the market reform that started in the 1980s, the three-tier referral system and the community healthcare services across the country gradually collapsed. The rollback of state investments in health left many small outpatient clinics to their own devices. They had to either close down or reduce their work to a few basic for-fee services (Blumenthal and Hsiao 2005). Of course, government funding was not completely withdrawn from healthcare provision, but it went mostly into building large secondary and tertiary care hospitals. With their cutting-edge medications, equipment, and procedures, these institutions in turn charged clients exorbitant fees. In the case of psychiatry, my fieldwork shows that family members were primarily responsible for hospitalizing patients and paying for their hospital bills. As a result, patients were constantly shuffled between psychiatric institutions and their homes (Ma 2014a; Phillips 1998).

Many psychiatrists whom I interviewed were unsatisfied with this concentration of mental health services in institutions. According to them, it denied treatment to patients in dire need of care, simply because the patients' families did not have the money for, access to, or knowledge of psychiatric hospitals. Such shortfalls in care not only limited the psychiatric profession's financial interests, but also went against the profession's ethical vision to "serve the people," which was a socialist legacy. Yan Jun, former director of the mental health division of the Ministry of Health and a key designer of the 686 Program, told me, "Community mental health is a long-cherished hope of psychiatrists from the older generation," who were trained prior to the market reform.

For psychiatrists, the SARS epidemic in 2003 brought a golden opportunity for change. The epidemic showed the neoliberal state the importance of public health for maintaining social stability and market productivity. Since then, much government money has been poured into reconstructing a public health system (Mason 2016). After a decade of development, this system now consists of a network of community health stations staffed by general practitioners and nurses. While most staffers still have to conduct fee-for-service work in order to generate income, they are also tasked and financially subsidized by municipal governments to provide basic public health services that range from immunization to HIV/AIDS control.

The state's renewed attention to public health was exciting, but at the beginning it was limited mainly to preventing and controlling infectious diseases. How could psychiatry share the spotlight and insert itself into this new wave of public health investment? After all, because mental illness was often perceived as "unreal" and lacking an effective cure, psychiatry was often despised by doctors in other medical specialties, and it had not fared well in attracting policy initiatives or government funding. Now that the government had shown more interest in health, leading psychiatrists felt it urgent to persuade the government and the general public that mental illness was not only real, but would also pose harm to both the individual body and the body politic.

By analyzing media reports and policy documents, I discovered that psychiatrists and allied policymakers have mobilized two geographically specific images of mental illness to achieve this end. One image depicts patients—mostly in rural areas—who are confined at home by their family members. In the late-nineteenth century, medical missionaries from the United States invoked this image of home confinement to portray the Chinese culture as barbaric and oppressive and to legitimize the asylums they established as efforts to liberate the insane individuals (Ma 2014b). In their attempt to construct a public mental health infrastructure, contemporary Chinese psychiatrists have once again highlighted home confinement, but not so much to criticize the Chinese culture as to demonstrate the lack of mental health knowledge and accessible psychiatric services in rural areas, as well as the tragic consequences that follow.[7] State investments, then, are seen as necessary for ending the humanitarian tragedies.

In response, the "Unlock Project" (*Jiesuo Gongcheng*) became one of the first components of the 686 Program. Launched in 2005, the project was aimed at discovering and freeing patients who had been locked up at home, offering them inpatient treatment, and ideally returning them to homes that were caring rather than constraining. A 2015 study, conducted by the project's leader, shows that it did reach hundreds of patients, improve their social functioning, and reduce their families' feelings of burden (Guan et al. 2015). However, a philanthropist who had worked on multiple "unlocking" efforts in rural Guangdong told me that there were only so many patients locked up in a certain area; after finding and "unlocking" them all, he and his psychiatrist friends did not know what to do next. He also lamented that without long-term support for them and their families, some unlocked patients were not welcome home, and still others had worsened conditions over time.[8] Partly due to these difficulties, unlocking has become less of a focus of the 686 Program, but the ideal of humanitarian care from the state is still very much alive.

Meanwhile, the image of home confinement of patients in rural China has been invoked by world-renowned psychiatrists to show the mental health disparities between low- or middle-income and high-income countries and to deplore the "failure of humanity" that puts individuals with mental illness in "the worst of moral conditions" (Kleinman 2009, 603). Those psychiatrists have thus advocated for "scal[ing] up the coverage of services for mental disorders . . . especially in low-income and middle-income countries" (Lancet Global Mental Health Group 2007, 1241). This global mental health agenda has in turn been taken up by Chinese psychiatrists in their call for more regular and comprehensive services for poor patients beyond mere unlocking.

Another image that has been mobilized to advocate for a community mental health infrastructure is that of mentally ill patients inflicting blind violence on strangers, especially in densely populated urban spaces.[9] As Robert Castel has pointed out, while such dangers are all-or-none occurrences that are ultimately unpredictable, psychiatry has turned them into risks that are inherent in every mentally ill patient (Castel 1991). According to policy documents of the 686 Program, psychosis distorts patients' mental activities from reality, directing them to injure and kill (e.g., Ministry of Health 2012a). Society therefore needs to be defended by having a public mental health infrastructure in place to detect those patients and to keep them under surveillance before they actually cause any harm. The ideas of risk and surveillance, then, construct "a vast hygienist utopia" that "plays on the alternate registers of fear and security" and aspire "for a life to which nothing happens" (Castel 1991, 289).

Besides—and intertwined with—this appeal to the security logic that is pertinent to the development of psychiatry in many places, the image of violence also elicits concerns specific to the postsocialist body politic. Since the late 1990s, with the rise of socioeconomic inequalities and popular unrest, the Chinese state has been increasingly preoccupied with any actual or potential threat to its governance and general social stability. A large amount of government funding has been channeled to "stability maintenance" (*weiwen*) work, and a cross-sector collaborative system has been built at every level of government (Lee and Zhang 2013). Several leading psychiatrists told me that in order to grasp the *weiwen* regime's attention, they and their colleagues had no choice but to highlight threats of violence that mental illnesses pose to social stability. The threats are purportedly graver in urban areas, because a violent act can cause more harm in a crowded setting, and because patients among the urban floating population are more difficult to discover and manage.

The images of rural tragedies and urban violence have thus played key roles in demonstrating the need to serve and surveil patients with severe mental illness. Together, they have constructed a dream for a caring, secure, and stable China, a national dream that is also supported by trends in global psychiatry. Compared to psychiatric hospitals, a community mental health infrastructure is more suited to fulfill this dream, because it can provide services and surveillance in situ. This vision has been well received by the state. Since 2006, a mental health joint conference (*lianxi huiyi*) mechanism has been established at all levels of government, involving not just the health sector but also more powerful government sectors like propaganda, finance, and public security in implementing community mental health. Moreover, state funding for the 686 Program rose from RMB 6.86 million (less than USD 1 million) in 2004 to RMB 473 million (about USD 73 million) in 2014, making it one of the biggest public health programs in the country.

Global Science, National Population

As I have mentioned, the 686 Program has two primary missions: providing service for mentally ill patients and ensuring surveillance of them. How do these two missions coexist? Is one just a cloak for the other, as suggested by the notion of "kindly power" (Yang 2015)? In this section, I will show that in the 686 Program, numbers have legitimated both service and surveillance, as well as allowing for commensuration between them.

As psychiatrists involved in the 686 Program told me, a question that they often face from colleagues is why community mental health was even necessary in China, given that there are still empty beds to fill in many psychiatric hospitals.[10] Moreover, even though home confinement and patient violence are captivating images, their actual occurrences are rare. Therefore, the construction of a nationwide community mental health infrastructure needs to be legitimized by the existence of many more patients who may have been neglected/abused or who may commit violence in the future, all of whom cannot be contained in institutions. Or in the words of Director Yan Jun, a large, untapped patient *population* has to be identified in order to demonstrate a hidden but objective *need* for service and surveillance, a need that is much broader than the expressed *demands*.

Epidemiology has provided the program's designers with a perfect tool to achieve this purpose. Until 2012 (Huang et al. 2019), there were only three transregional psychiatric epidemiological surveys in China, in the years of 1982 (Shen, Chen, and Zhang 1986), 1993 (Zhang et al. 1998), and 2001–2005 (Phillips et al. 2009), respectively. Each survey covered urban and rural sites in four to twelve provinces. These surveys all used globally circulating disease categorizations (such as the Diagnostic and Statistical Manual of Mental Disorders, fourth edition, in the 2001–2005 survey) and measurements, thereby treating mental illnesses as universal entities (Rosenberg 2007). While all three surveys compared disease prevalence in urban and rural areas, only the 1982 one analyzed differences among provinces and tried to connect them to local cultural, economic, and geographical conditions. In contrast, the two more recent surveys simply took data collected from specific sites as estimates for the whole country. In any case, across the three surveys, the estimated national prevalence of schizophrenia increased only slowly over time (5.69 per thousand in 1982, 6.55 per thousand in 1982, 10 per thousand in 2001–2005). Meanwhile, the prevalence of mental illness in general rose dramatically from 13.47 per thousand in 1993 to 175 per thousand in 2001–2005, mostly due to the rise of depression and anxiety disorders. Besides the fact that the tools used in the most recent survey were more sensitive to detecting depression, one might also attribute this drastic change to the increased popularity of depression as a framework for emotional experience, expression, and diagnosis (Lee 1999).

In his analysis of disability politics in postsocialist China, Matthew Kohrman discovered that with its scientific aura, the globally circulating prevalence rate of disability "took on tremendous normative authority" when Chinese officials sought to codify and quantify disability (Kohrman 2003, 14). Similarly, in the case of mental illness, the prevalence rates identified in the most recent survey are comparable to those in Euro-American countries. For example, Dr. Yu Xin explained to journalists that the 1 percent rate of schizophrenia in China matched the international rate (Han and Ren 2013). He and other psychiatrists have thus taken these rates as reliable estimates of the mentally ill population in China.

Interestingly, although depression has become the most prevalent mental illness and alarming to the general public, the 686 Program does not include it as a targeted illness. The program's leaders told me that there are simply too many depressed patients for the state to care for. They also admitted that the security state, to which they appeal for resources, is more—if not only—concerned with people who may harm others, while patients with depression typically harm only themselves. In fact, the conditions that the 686 Program covers with the umbrella term "serious mental illness"—schizophrenia, schizoaffective disorder, bipolar disorder, paranoid disorder, mental retardation with psychosis, and epilepsy with psychosis—all have a strong psychotic element (Ministry of Health 2012a). As psychiatrists and government authorities see it, psychosis indexes potentials of violence that need to be constantly managed. It also exposes patients to substantial disability, vulnerability, and stigmatization.[11] Therefore, the state's medical humanitarianism and security management are commensurated in the categorization and prevalence measurement of psychosis/serious mental illness. With its stamp of global validation, the 1 percent prevalence rate of schizophrenia—or 16 million patients, if counting all serious mental illnesses—now indisputably indicates a large population that is simultaneously vulnerable and potentially violent. Given that existing hospital beds can cover 10 percent of this population at most,[12] the prevalence rate also indicates enormous needs for service and surveillance hidden outside the institutions, which the 686 Program is built to satisfy.

Numbers and the Unfolding of the Community Mental Health Infrastructure

Now that the target population of community mental health interventions has been identified, its members need to be found, located, and counted. As David Bray has pointed out, "community" in contemporary China is envisioned a basic unit of governance, which has "a distinct territory" and is "run by a team of officials employing a standardized repertoire of bureaucratic procedures" (Bray 2006, 535). At the local level, the community mental health infrastructure is composed of general practitioners assigned by community health stations to conduct mental health work. After receiving basic training in psychopathology, those CMHPs work on detecting, serving, and monitoring patients in their jurisdictions. According to principles of scientific management, it is only after the number of patients in a community is known that the number of CMHPs assigned there can be determined, along with their required workload. Meanwhile, if discovering a patient is the precondition of serving and surveilling him, then the number of patients detected can indicate the quality of a CMHP's work and the level of attention a community administration devotes to scale up mental health. As the director of the 686 Program in Nanhua told me, "The state has to use these indicators to see if you have provided services." Therefore, rendering the patient population visible and enumerating patients across space have become essential for the community mental health infrastructure to unfold.

In 2012, the Ministry of Health issued an "Assessment and Evaluation Plan for the Management and Treatment Work for Serious Mental Illnesses." The plan held the provincial departments of health accountable for a range of tasks, including detecting a certain amount of seriously mentally ill patients and registering their information in an online database (Ministry of Health 2012b). As the program's designers told me, the 1 percent prevalence rate of schizophrenia indicates the number of patients there are in the country, the ideal service and surveillance capacity of the 686 Program, and the number of patients that program workers ultimately should be able to discover. Moreover, measured with a universal disease categorization, calculated by averaging across the nation and validated by a global uniformity, the 1 percent prevalence rate has allowed psychiatrists to see China not only as equivalent to the external world in terms of illness onset, but also as an internally even landscape of psychopathology. In the words of a prominent psychiatrist in Guangdong, "Illness occurrence knows no regional difference." The number 1 percent, then, has been set as the ultimate ideal detection rate—the number of patients registered in the database divided by the size of the resident population—for every province.

Although patients with serious mental illness are seen as evenly distributed across the country, and although the ultimate goal of infrastructural development is the same everywhere, the current status of psychiatric services varies geographically. As designers of the 686 Program saw it, a region's capacity in patient detection and community infrastructural development is shaped primarily by the existing capacity of psychiatric institutions and personnel there, and the latter is in turn determined by the region's economic development. In postsocialist China, discursive and material forces of market reform have aligned geographical regions—West, Central, East—along a linear trajectory of economic development. Therefore, in order to make the 686 Program more realistic, its designers decided to push it forward with different intensities in different geo-economic regions. The 2012 assessment and evaluation plan required that from 2012 to 2015, the target detection rate for each province was to gradually rise from 2.5 per thousand to 4 per thousand. Every year, the program's implementation in each province would be evaluated by the national overseers, with 20 points out of 1,000 assigned to evaluate whether the detection target was met. For each 0.5 per thousand under target, 4 points would be deducted for an Eastern province; 3 for a Central province; and 2 for a Western province. Granted, 20 points out of 1,000 might not seem like much, and the assessment and evaluation plan included many other quantitative targets. However, for many program officials, the discovery and registration of patients were the first steps toward serving and managing them, so the detection rate was the most important among all targets. In any case, we can see that the target detection rate of the 686 Program directs different regions to march at different paces toward the same goal of infrastructural development, which will match the homogenous landscape of mental illness and service/surveillance needs across the country.

In the logic of the program's designers, program targets such as the detection rate are both validated by global epidemiology and tempered by national reality. They thus provide "the source of *truth*" (Greenhalgh 2005a) to guide the construction of the community mental health infrastructure, as well as the pursuit of a caring, secure, and stable state. A closer look at the numbers' assumptions and operations, however, reveals how controversies around the program may have arisen, such as the accusation of "quota apportioning." For one, the *number* of patients in a given territory—especially the magical 1 percent—is made to work in several modalities at once: it indicates patients who are already out there waiting to be discovered, anticipates or helps to plan the size of the infrastructure to be constructed, and evaluates the actual workings of the infrastructure. As seen in the Zhengzhou incident, this mixing of the indicative, anticipatory, and evaluative modes of enumeration in the patient detection rate has evoked in many people the nightmare of socialist central planning.

Many Chinese people remember that in the era of high socialism, the state used administrative means to assign production quotas to grassroots units and to evaluate people's work accordingly, rather than allowing actual market demands and production capacities to determine production goals. In their desperate attempts to meet the unduly high quotas, local people often had to exaggerate the quantities of their products and then sacrifice their means of subsistence to make up the difference. The upshot was widespread hunger and death from 1959 to 1961 (Feng 2014). In the reform era, many social and economic planning processes are still target driven,[13] except that technocrats have increasingly sought to ground the targets in science. For example, Susan Greenhalgh has discovered that in the late 1970s, Chinese missile scientists applied cybernetics to produce a striking prediction of population growth, which led to the making of the radical one-child policy (Greenhalgh 2005b) In this way, "scientific management . . . combined with socialist target setting" to create ever more powerful numbers and demanding tasks (Greenhalgh 2005a, 363). Accordingly, we can understand why, in the Zhengzhou incident, critics like Huang Xuetao were not satisfied with psychiatrists' use of epidemiological data to claim facticity for the patient detection rates. After all, it was the numbers' seamless movement from scientific estimates to planned quotas/targets and evaluative standards that had created the historical traumas.

Another source of controversies lies in how these numbers circulate. We have seen that in the Zhengzhou incident, CMHPs complained about the heavy pressure they faced to fulfill the 2 per thousand detection rate. In response, officials from the Ministry of Health claimed that the detection rates were merely guidelines for local work, not hard-and-fast quotas. This was an honest defense. After all, the 2012 assessment and evaluation plan held only the provincial governments accountable for meeting the detection rate, not the lower-level units. In my interview with her, Director Yan Jun used basic statistical rules of epidemiology to explain why it was the case: a statistical estimate such as a disease prevalence rate is more likely to hold stable when the base population is large, but it gets more unstable when the base gets smaller; ergo, while it shouldn't be a problem to require a province, or even a big city, to detect a certain proportion of seriously mentally ill patients among its population, it would be problematic to require a small district or a smaller community to do so. The all-too-mechanical (*tai jixie de*) management style of the low-level bureaucrats, Yan contended, was to blame for carrying the provincial-level detection rate all the way down to the communities.

However, by blaming the "number pressure" on human agents, Yan and her Beijing colleagues failed to see the power that numbers wield in assembling an infrastructure. Epidemiological statistics, as Yan explained, do allow for local variations. Yet they also render those variations marginal, merely a supplement to the nation-wide means of disease prevalence, and not in need of further investigations or bureaucratic attention. The detection rate that was designed based on epidemiological statistics has thereby assumed a "black box" quality.[14] That is, it contains, commensurates, and conceals not just different dreams of community mental health reconstruction and different modes of numerical governance, but also diversities in mental distress and service needs. This black-boxed number is able to travel up and down as what Bruno Latour has called an "immutable mobile," that is, "a technoscientific form that can be decontextualized and recontextualized, abstracted, transported, and reterritorialized, and is designed to produce functionally comparable results in disparate domains" (quoted in Collier and Ong 2005, 11).

The downward travel of the black-boxed, immutably mobile numbers has been facilitated by the existing bureaucratic pathways. The postsocialist Chinese government has been run on a cadre accountability system, which allows higher-level units to designate policy targets for units one level below and to assess the latter's work accordingly. Because the assessment results can determine officials' promotion or demotion, scholars have found that the policy targets loom large for officials, who often pass down this pressure to their even lower-level counterparts (O'Brien and Li 1999; Zhao 2007). As a government sponsored infrastructure, the 686 Program has been explicitly modeled after this accountability system. Under the 2012 national assessment and evaluation plan, provincial program leaders enjoy the freedom to design assessment and evaluation plans for subordinate cities. Yet they also feel the pressure to achieve the goals set by the national plan. Many provincial program leaders have thus chosen to pass down the numeric targets they had received, albeit with some small tinkering considering each city's organizational capacity. The same is true for municipal program leaders in their plans for the districts. In 2013, for example, although the national target detection rate was 3 per thousand, the municipal program leaders in Zhengzhou lowered the rate for its districts to 2 per thousand, because the 686 Program had just begun there. In contrast, because the program had started in Guangdong relatively early, its provincial leaders raised the target rate for most cities to 4 per thousand; in economically developed cities like Shenzhen, the target rate was even as high as 6 per thousand. This was to make sure that the province could reach the ultimate goal of 1 percent sooner. In almost every site of program implementation, the degree to which the target detection rate and other numerical goals are met matters greatly to one's institutional standing and year-end bonus. As such, they preoccupy everyone in the program.

Therefore, once designed and pushed forward by experts and policymakers at the central level, the detection rate and other program targets have gained a certain degree of inertia, moving and reproducing themselves automatically along the bureaucratic pathways. In their travels, these numerical black boxes have also carved out communities as grid-like territories of psychopathology and of mental health service and surveillance. In other words, they have become the building blocks of the community mental health infrastructure. Yet for many CMHPs whom I encountered, these numerical black boxes carry with them unbearable bureaucratic pressure, in response to which secretive tactics have to be used. It is this interplay between centralized measures and localized countermeasures that has evoked people's historical anxiety with quota apportioning.

Uneven Terrains of Numerical Governance

Whether taken as a dream of modernized, globalized development or as a nightmare of socialist planning, numbers in the 686 Program have assumed a "black box" quality. The previous sections have shown how the black box was made and how it has been enabled to travel. In this section, I examine the actual circulation of these numbers in everyday program practices, focusing on what it includes and what it excludes, when it works and when it breaks down.

By setting up a series of quantitative targets, especially targets in detecting and registering patients, the 686 Program has built a large mental health infrastructure from an almost blank slate. In Nanhua, where I conducted most of my fieldwork, the program started by having general practitioners work part-time on mental health. "Part-time" here could mean spending as little as half a day each week, as those general practitioners were typically assigned multiple lines of public health work, all of which were underfunded. In order to support themselves, they had to devote most of their time to providing for-fee services. Gradually, as the tasks of patient detection, service, and surveillance have become emphasized and systematized, these general practitioners have received more financial and organizational support for mental health work. A growing number of them have become full-time CMHPs. In turn, the procedure to guide and evaluate their work has become increasingly elaborate, encompassing more and more quantitative targets.

The pressure of quantification has produced a "target obsession and numbers mania" (Greenhalgh 2005a, 368), which drives CMHPs not only to see patients as numbers, but also to devise tactics of inclusion and exclusion. The 686 Program follows the principle of dynamic management, which requires CMHPs to update the patient database and adjust their intervention plan whenever a patient's status changes. In Nanhua, however, I learned that some CMHPs deliberately kept the files of dead patients on record. Some other CMHPs did not change a patient's address long after he or she moved out of their jurisdiction. In certain communities where big psychiatric institutions were located, the CMHPs would include all the inpatients there in their database, despite the fact that they did not provide any "community" services to those inpatients, who, moreover, came from all around the city or even nearby areas. All these were done just to keep the numbers of detected patients high.

However, having too many patients on file can bring its own problems. Alongside the detection rate, the 686 Program also requires that a certain proportion of patients detected—80 percent in Nanhua—be managed by CMHPs. That is, CMHPs have to regularly visit patients' homes to collect information on their symptom fluctuations, medication compliance, and risk levels, as well as providing patients and their families with information on relevant welfare policies.[15] In my fieldwork, many CMHPs complained that it was impossible to do all the work: given the population density in Nanhua, meeting the target detection rate could mean having as many as a few hundred patients at hand. Moreover, CMHPs' work has been subjected to intensified monitoring, as program leaders seek to allay public concerns with data fabrication and fear of "quota apportioning." In Nanhua, municipal program leaders visit community health stations to check the records every six months. There, in front of the CMHPs, they make "cold calls" to patients registered in the database to see whether the information is real, whether the CMHPs have indeed paid the patients regular visits, and so on. If one case proves to be fake or not properly managed, all the points in the corresponding category may be deducted. In effect, municipal program leaders perform a "ritual of verification" (Power 1997) to highlight the importance of truth in data collection and to legitimate the data collected.

Here I offer an enactment of this ritual or verification. During a program evaluation trip to the most populous district of Nanhua, Dr. Dong, a municipal supervisor, cold-called a patient on file and asked him if he had been visited by any doctor, was taking any medication, or had received any other services. The patient said no to all those questions. After she hung up the phone, Dr. Dong turned to the CMHP Dr. Liang, who was standing nearby and shaking visibly:

> DONG: You've never visited the patient, have you?
> LIANG: Well, maybe I've called him . . . and told him to come to the psychoeducation lectures.
> DONG: Did he come? Did you actually see him? Did you introduce yourself?
> LIANG: I don't remember.
> DONG: So do you really know his conditions?
> LIANG: How could I know his conditions when I couldn't prescribe any medications?
> DONG: You could just ask. Anyway, you should have at least talked to him in person, no matter where it was.
> LIANG: (forcing a smile) I'm sorry. I didn't know that the evaluation had become so thorough. . . . Do you think the [target detention and management] rates will get higher and higher?

Here, we can see that the ritual of verification puts all the blame on CMHPs and frames them as potential liars, while ignoring the structural constraints that make their work difficult. Besides their heavy workload—Dr. Liang was in charge of 280 seriously mentally ill patients in a community of 45,000 residents—another challenge suggested by Dr. Liang is the limited capacity of CMHPs to provide services to patients and families. Most CMHPs have not received in-depth training in psychiatry, and according to the current regulations, they are not allowed to make diagnoses or prescribe psychopharmaceuticals. The best they can do is to provide seriously mentally ill patients and their families with healthcare and policy information, arrange for impoverished patients to receive free but rudimentary medications from designated psychiatrists, perform annual physical examinations on patients, and refer them to inpatient treatment if necessary. Therefore, in metropolitan areas like Nanhua, where institutional psychiatry is relatively well established, community mental health services so far seem crude, even for residents who may not be able to afford specialized institutional services. The limitation lies not only in the CMHPs' skillsets, but also in the fragmented and stringent healthcare regime. As one CMHP put it, "What good would it do if the annual physicals showed something wrong [other than mental illness] but we couldn't give the person free treatment? Of course people won't come!"

Meanwhile, community or outpatient services for some other psychiatric categories—such as depression—are in high demand, but the way in which program targets have been set prevents CMHPs from addressing these popular concerns. Ironically, some CMHPs whom I interviewed joked about sending all the depressed patients in their communities for re-diagnosis to see if some would meet the (loosened) criteria for bipolar disorder. That way, not only would they have a better chance in meeting the target detection rate, but those depressed patients could also benefit from the program's services. No one I knew was serious about carrying out this plan, for fear of the legal repercussions of a misdiagnosis. Yet as Paul Brodwin argues in his study of American community psychiatry, "What people [frontline caseworkers] say about their jobs reveals the rough edges of today's community psychiatry. Their ethical comments in particular illuminate some of the deepest and most intractable problems" (Brodwin 2013, 4). In this case, the problem revealed by the CMHPs' comments is the 686 Program's preoccupation with certain numerical targets, its narrow purview of mental illness, and its skewed orientation of service provision.

Remember that for its designers and state sponsors, the 686 Program, by simultaneously serving *and* surveilling seriously mentally ill patients, enables the state to enact humanitarian care *and* security management, to exert a genuinely kind power. However, given how numbers shape and circumscribe the work of CMHPs, the two missions of community mental health sometimes threaten to disentangle. My fieldwork shows that through personal experience with or hearsay of the program's work, urban residents, especially (suspected) patients and their family members, feel the state's detective gaze and its desire to obtain information on them. This feeling suggests to them not the state's intention or capacity to serve, but its anxiety to make sure that they do not stir up trouble. They are concerned about how far their information will travel beyond their own knowledge, as well as any negative consequences of being in the "system." This apprehension is especially preponderant for young and/or working people who have a strong sense of privacy and who might lose their jobs or future to the stigma of mental illness. Therefore, they often refuse to work with CMHPs or to provide any personal information. In Nanhua, this has become a particularly thorny problem in a district populated with college students and middle-class professionals. In 2013, when the city's target detection rate of seriously mentally ill patients was 4 per thousand, the district's actual detection rate was lower than 1 per thousand. During a municipal assessment tour that I followed, some CMHPs complained about how they were often denied entrance or even chased away when they tried to visit people reported to have serious mental illness. Because the district, thanks to its low detection rate, had been repeatedly called out by the provincial program leaders as a negative example, a local official even grumbled, "They [the leaders] simply want our district to go crazy!"

Finally, while target obsession may have driven CMHPs to include people who should not have been, it has also selectively excluded those in need of service, especially city migrants. Without household registration status (*hukou*) in the city, migrants are not entitled to most community mental health services, however limited they may be. Besides, CMHPs often find it difficult to track down and follow up with migrants. Given that CMHPs have to hit a high rate of patient management, they tend to ignore migrants in their everyday work. The only exception is when occasionally a migrant becomes a mad assailant committing acts of violence in the public; in that case, the CMHP in charge of the area will be held accountable for not having detected and properly managed that person. This unmanaged/unmanageable and disruptive appearance of migrants in turn reinforces the image of violent psychotic patients roaming free in the city, tilting the balance of community mental health further toward security management.

The selective exclusion of migrants has troubled the 686 Program's developmentalist vision. As previously mentioned, the program's designers expect that the more economically prosperous a region is, the easier it is to detect patients there. Yet in Guangdong at least, the situation has proven to be the opposite: in 2013, cities that were economically less prosperous and that had a trend of outmigration scored higher rates of patient detection, some even as high as 10 per thousand, while Nanhua, a booming metropolis full of migrants, was struggling to reach 4 per thousand. Program leaders and CMHPs in Nanhua blamed this difficulty on the "unfair" way in which the detection rate was calculated: while the denominator was the number of all permanent residents in an area—that is, anyone who had lived there for more than six months—the numerator, in reality, could only be the number of patients with household registration status. It was not surprising, then, that cities with populations flowing out had an easier job meeting the target detection rate than cities with large influxes of migrants. This analysis makes sense. However, instead of discussing what could have been done to include migrants in the mental health service provision at their new homes, several CMHPs proposed to drive them out of the equation—that is, not to count them in the city's base population. Here, we can see how a pursuit of "pure" and "correct" numbers may lead to infrastructural exclusion.

Conclusion

I argue in this chapter that numbers are central to the emergence and everyday operations of community mental health in China. Validated by global science and tempered by national and local interests, numbers have allowed different dreams of governance—service and surveillance, humanitarian care and security management—to commensurate, to gain the state's recognition, and to identify a target population for intervention. Operating simultaneously as epidemiological estimates, program targets, and evaluation standards, numbers have become basic building blocks of the community mental health infrastructure. Their circulation along the existing bureaucratic pathways has carved out communities as grid-like territories, where patients are distributed and practitioners are summoned. Yet it is also in their circulation that numbers exert quota-like pressure on practitioners, subject them to impossible tasks and blame, outrun other aspects of the program, unhinge service from surveillance, and exclude certain groups from communities. These effects of numerical governance have thus turned the dream of community mental health into a nightmare for some people and produced unkind consequences of a psychiatric power that is genuinely intended to be kind.

The 686 Program not only is an endeavor to construct a nation-wide community mental health infrastructure in China, but has also become "the world's largest—and arguably the most important—mental health services demonstration project" (Good and Good 2012). As such, its practices can shed light on some key dynamics in global mental health: how numbers standardize distress and simplify psychiatric expertise, shifting it to the shoulders of primary care practitioners (Brodwin 2013); how they scale up the need for mental health services and scale down interventions to the local level (Bemme and D'souza 2014); and how they commensurate different forces of governance or create frictions between them (Tsing 2011). Moreover, many programs of community governance in China rely on numbers in their daily operations. The case of the 686 Program thus suggests that by tracing the genealogy of the numbers' making and the trajectory of their circulation, we can gain a better understanding of how different aspects of community governance may coexist or disentangle, and whether or how these various "Chinese dreams" may live out.

NOTES

Research for this chapter was supported by the New Generation China Scholar Fellowship and co-sponsored by the Ford Foundation and the University of Chicago Center in Beijing. Comments on previous drafts were generously provided by Stephen Baker, Susan Greenhalgh, Lisa Hoffman, Kevin Ko, Scott Selberg, Ana Vinea, Li Zhang, participants of the Harvard Fairbank Center workshop "A Better Life Through Science and Biomedicine?," and 2015–2016 fellows of the University of Chicago Urban Network. I also thank the two anonymous reviewers for their remarks on the first submitted draft.

1. In policy documents, the full name of these personnel is "community doctors for preventing and treating mental illness" (*shequ jingshenbing fangzhi yisheng*).

2. For the name of everyone in my fieldwork, I follow the Chinese custom of putting their family name first.

3. Espeland and Stevens define commensuration as "the expression or measurement of characteristics normally represented by different units according to a common metric" (Espeland and Stevens 1998, 315).

4. In her analysis of the indicator culture in the human rights world, Sally Engle Merry (2016) discovered that indicators entail two forms of inertia: "expertise inertia," which means that "insiders with skills and experience have a greater say in developing measurement systems than those without" (6); and "data inertia," which refers to "the fact that existing data determine what an indicator can measure" (7). My use of the term "inertia" echoes these meanings, but my emphasis is more on how program targets, once determined, have the power to move automatically and in largely unchanged forms.

5. Theodore Porter has suggested that "numbers are the medium through which dissimilar desires, needs, and expectations are somehow made commensurable" (Porter 1996, 86).

6. Nanhua is a pseudonym. I use it in order to protect my interlocutors from potential institutional repercussions. For the same reason, I deliberately keep the city's demographic information vague. Note that every city in China has administratively designated urban and rural areas, although most people in Nanhua live in urban areas. Therefore, what I describe later on as urban and rural imaginations of community mental health apply to Nanhua as well.

7. Such images abound in media reports—e.g., Kong and Li 2013; LaFraniere 2010. Because these reports on home confinement often end with psychiatrists going in and unlocking the patients or appealing for government-funded medical intervention, they were likely to be orchestrated by the psychiatrists themselves.

8. Anthropologists have reminded us that the temporality of crisis and rescue makes it difficult, if not impossible, for humanitarian projects to be transformed into permanent infrastructures (Redfield 2005).

9. For psychiatrists' discussions of patient violence in the press, see Cao and Xu, 2013; Cheng 2015; He and Li 2015; Zhou 2010.

10. Official statistics show that as of 2006, only 87 percent of all psychiatric beds in China were occupied (CDC 2008).

11. This view can also be found in the global mental health community. See Lancet Global Mental Health Group, 2007.

12. According to a survey conducted by the Ministry of Health, as of 2005, there were on average 1.12 psychiatric beds per 10,000 individuals across China, and the average course of an inpatient stay was forty-six days (CDC 2008).

13. Li Zhang has used the term "late socialism" to designate the period after the Cultural Revolution and before the shift to the Jiang Zemin leadership in 1993 (Zhang 2012, 16).

14. Bruno Latour has pointed out that at the heart of all scientific production is the process of blackboxing, that is, "the way scientific and technical work is made invisible by its own success" (Latour 1999, 304).

15. In March 2017, the Ministry of Health published the latest *Code of National Public Health Service*. It eliminated the management rate of seriously mentally ill patients from service evaluation indicators, along with a few other public health targets (Ministry of Health 2017). A public health official in Shaanxi remarked that this was to shift the field's focus from quantity to quality of service (Wang 2017). Given that all the other community mental health targets have remained in place, I doubt that radical changes will take place soon.

References

Adams, Vincanne. 2016. *Metrics: What Counts in Global Health*. Durham, NC: Duke University Press.

Bemme, Doerte, and Nicole A D'souza. 2014. "Global Mental Health and Its Discontents: An Inquiry into the Making of Global and Local Scale." *Transcultural Psychiatry* 51(6): 850–874.

Blumenthal, David, and William Hsiao. 2005. "Privatization and Its Discontents: The Evolving Chinese Health Care System." *New England Journal of Medicine* 353(11): 1165–1170.

Bray, David. 2006. "Building 'Community': New Strategies of Governance in Urban China." *Economy and Society* 35(4): 530–549.

Brodwin, Paul. 2013. *Everyday Ethics: Voices from the Front Line of Community Psychiatry*. Berkeley, CA: University of California Press.

Cao, Xindong, and Qiuyue Xu. 2013. "Incidents and Accidents of Violence in Serious Mental Illnesses: An Analysis of the Current Situation and Cause." *China Medical Tribune*. http://news.medlive.cn/all/info-progress/show-48065_60.html.

Castel, Robert. 1991. "From Dangerousness to Risk." In *The Foucault Effect: Studies in Governmentality*, edited by Graham Burchell, Colin Gordon and Peter Miller, 281–298. Chicago: University of Chicago Press.

CDC—*See* Chinese Center for Disease Control and Prevention.

Cheng, Jingwei. 2015. "In Guangzhou, 22 Cases of Psychiatric Patient Violence Were Handled in the First Nine Months of 2015." *China News*. http://www.chinanews.com/sh/2015/10-20/7580066.shtml.

Chinese Center for Disease Control and Prevention, ed. 2008. *Collection of Reports on Mental Health Policy Research*. Beijing: People's Health Press.

Cho, Mun Young. 2010. "On the Edge between 'The People' and 'The Population': Ethnographic Research on the Minimum Livelihood Guarantee." *China Quarterly* 201(1): 20–37.

Cho, Mun Young. 2013. *The Specter of "The People": Urban Poverty in Northeast China*. Ithaca, NY: Cornell University Press.

Collier, Stephen J., and Aihwa Ong. 2005. "Global Assemblages, Anthropological Problems." In *Global Assemblages: Technology, Politics, and Ethics as Anthropological Problems*, edited by Aihwa Ong and Stephen Collier, 3–21. Malden, MA: Blackwell Publishing.

Espeland, Wendy Nelson, and Mitchell L. Stevens. 1998. "Commensuration as a Social Process." *Annual Review of Sociology* 24(1): 313–343.

Feng, Xiaocai. 2014. "Discount and Exaggeration: Two Tendencies in Quota Politics." *Open Times*. http://www.opentimes.cn/bencandy.php?fid=373&aid=1782.

Good, Byron J., and Mary-Jo DelVecchio Good. 2012. "Significance of the 686 Program for China and for Global Mental Health." *Shanghai Archives of Psychiatry* 24(3): 175–177.

Greenhalgh, Susan. 2005a. Globalization and Population Governance in China. In *Global Assemblages: Technology, Politics, And Ethics as Anthropological Problems*, edited by Aihwa Ong and Stephen Collier, 354–372. Malden, MA: Blackwell Publishing.

Greenhalgh, Susan. 2005b. "Missile Science, Population Science: The Origins of China's One-Child Policy." *China Quarterly* 182(2): 253–276.

Greenhalgh, Susan. 2008. *Just One Child: Science and Policy in Deng's China*. Berkeley: University of California Press.

Greenhalgh, Susan, and Edwin A Winckler. 2005. *Governing China's Population: From Leninist to Neoliberal Biopolitics*. Palo Alto, CA: Stanford University Press.

Guan, Lili, Jin Liu, Xia Min Wu, Dafang Chen, Xun Wang, Ning Ma, Yan Wang, Byron Good, Hong Ma, and Xin Yu. 2015. "Unlocking Patients with Mental Disorders Who Were in Restraints at Home: A National Follow-Up Study of China's New Public Mental Health Initiatives." *PLoS ONE* 10(4): e0121425. doi:10.1371/journal.pone.0121425.

Han, Xiaomei, and Bo Ren. 2013. "Experts Claim that the Detection Rate of Psychiatric Patients Is Far Lower than the Prevalence Rate." *Caixin*, October 10. http://china.caixin.com/2013-10-10/100589837.html.

He, Dayuan, and Rong Li. 2015. "Shenzhen Uses Multiple Measures to Strengthen Comprehensive Service and Management of Patients with Serious Mental Illnesses." *Southern Metropolis Daily*, November 26. http://epaper.oeeee.com/epaper/H/html/2015-11/26/content_15103.htm.

Heberer, Thomas, and Christian Göbel. 2011. *The Politics of Community Building in Urban China*. New York: Routledge.

Huang, Xuetao. 2013. "Remarks on World Mental Health Day: 'Apportioning Quotas of Mental Illness' Is a Monster Born by Turning Medicine into Administrative Matters." *Lawyer Huang Xuetao's Blog*, October 11. http://blog.sina.com.cn/s/blog_608a961a0101dm6m.html.

Huang, Yueqin, Yu Wang, Hong Wang, Zhaorui Liu, Xin Yu, Jie Yan, Yaqin Yu, et al. 2019. "Prevalence of Mental Disorders in China: A Cross-Sectional Epidemiological Study." *The Lancet Psychiatry* 6(3): 211–224.

Kipnis, Andrew B. 2008. "Audit Cultures: Neoliberal Governmentality, Socialist Legacy, or Technologies of Governing?" *American Ethnologist* 35(2): 275–289.

Kleinman, Arthur. 2009. "Global Mental Health: A Failure of Humanity." *The Lancet* 374(9690): 603–604.

Kohrman, Matthew. 2003. "Why Am I Not Disabled? Making State Subjects, Making Statistics in Post-Mao China." *Medical Anthropology Quarterly* 17(1): 5–24.

Kong, Pu, and Tianyu Li. 2013. "'Men in Cages,' Special Report." *Beijing News*, July 11. from http://www.bjnews.com.cn/feature/2013/07/11/272800.html.

LaFraniere, Sharon. 2010. "Life in Shadows for Mentally Ill in China." *New York Times*, November 10.

Lancet Global Mental Health Group. 2007. "Scale Up Services for Mental Disorders: A Call for Action." *The Lancet* 370(9594): 1241–1252.

Latour, Bruno. 1999. *Pandora's Hope: Essays on the Reality of Science Studies*. Cambridge, MA: Harvard University Press.

Lee, Ching Kwan, and Yonghong Zhang. 2013. "The Power of Instability: Unraveling the Microfoundations of Bargained Authoritarianism in China." *American Journal of Sociology* 118(6): 1475–1508.

Lee, Sing. 1999. "Diagnosis Postponed: Shenjing Shuairuo and the Transformation of Psychiatry in Post-Mao China." *Culture, Medicine and Psychiatry* 23(3): 349–380.

Ma, Hong, Jin Liu, Yanling He, Bin Xie, Yifeng Xu, Wei Hao, Hongyu Tang, Mingyuan Zhang, and Xin Yu. 2011. "An Important Pathway of Mental Health Service Reform in China: Introduction of 686 Program." *Chinese Mental Health Journal* 25(10): 725–728.

Ma, Zhiying. 2014a. "Intimate Politics of Life: The Family Subject of Rights/Responsibilities and Mental Health Legislation." *Thinking* 40(3): 42–49.

Ma, Zhiying. 2014b. "An 'Iron Cage' of Civilization? Missionary Psychiatry, Chinese Family, and a Colonial Dialectic of Enlightenment." In *History of Psychiatry in Chinese East Asia*, edited by Howard Chiang, 91–110. London: Pickering and Chatto.

Mason, Katherine. 2016. *Infectious Change: Reinventing Chinese Public Health after an Epidemic*. Palo Alto, CA: Stanford University Press.

Merry, Sally Engle. 2016. *The Seductions of Quantification: Measuring Human Rights, Gender Violence, and Sex Trafficking*. Chicago: University of Chicago Press.

Ministry of Health of the People's Republic of China. 2012a. *Rules for Management and Treatment of Serious Mental Illnesses*. Beijing: Ministry of Health General Office.

Ministry of Health of the People's Republic of China. 2012b. *Assessment and Evaluation Plan for the Management and Treatment Work for Serious Mental Illnesses*. Beijing: Ministry of Health General Office.

Ministry of Health of the People's Republic of China. 2017. *Code of National Public Health Service*. Beijing: Ministry of Health General Office.

O'Brien, Kevin J., and Lianjiang Li. 1999. "Selective Policy Implementation in Rural China." *Comparative Politics* 31(2): 167–186.

Pearson, Veronica. 1995. *Mental Health Care in China: State Policies, Professional Services and Family Responsibilities*. London: Gaskell.

Peking University Sixth Hospital. 2012. "From 6.86 Million to 93.87 Million." http://www.pkuh6.cn/News/Articles/Index/101911.

Phillips, Michael. 1998. "The Transformation of China's Mental Health Services." *China Journal* 39(1): 1–36.

Phillips, Michael, Jingxuan Zhang, Qichang Shi, Zhiqiang Song, Zhijie Ding, Shutao Pang, Xianyun Li, Yali Zhang, and Zhiqing Wang. 2009. "Prevalence, Treatment, and Associated Disability of Mental Disorders in Four Provinces in China during 2001–05: An Epidemiological Survey." *The Lancet* 373(9680): 2041–2053.

PKU6—*See* Peking University Sixth Hospital.

Porter, Theodore M. 1996. *Trust in Numbers: The Pursuit of Objectivity in Science and Public Life*. Princeton, NJ: Princeton University Press.

Power, Michael. 1997. *The Audit Society: Rituals of Verification*. Oxford, UK: Oxford University Press.

Read, Benjamin. 2012. *Roots of the State: Neighborhood Organization and Social Networks in Beijing and Taipei*. Palo Alto, CA: Stanford University Press.

Redfield, Peter. 2005. "Doctors, Borders, and Life in Crisis." *Cultural Anthropology* 20(3): 328–361.

Rose, Nikolas. 1996. "The Death of the Social? Re-Figuring the Territory of Government." *Economy and Society* 25(3): 327–356.

Rosenberg, Charles. 2007. *Our Present Complaint: American Medicine, Then and Now.* Baltimore, MD: Johns Hopkins University Press.

Shen, Yucun, Changhui Chen, and Weixi Zhang. 1986. "Methodology and Data Analysis of the 12-Region Mental Disorder Epidemiological Survey." *Chinese Journal of Nervous and Mental Diseases* 19(2): 65–69.

Strathern, Marilyn. 2000. *Audit Cultures: Anthropological Studies in Accountability, Ethics, and the Academy.* London: Routledge.

Tomba, Luigi. 2014. *The Government Next Door: Neighborhood Politics in Urban China.* Ithaca, NY: Cornell University Press.

Tsing, Anna. 2011. *Friction: An Ethnography of Global Connection.* Princeton, NJ: Princeton University Press.

Wang, Hui. 2008. *Depoliticized Politics: The End of the Short 20th Century and the 1990s.* Beijing: SDX Joint Publishing Company.

Wang, Hui. 2017. "The State Announced Radical Changes in Public Health." *China County Health*, March 30. http://mp.weixin.qq.com/s?__biz=MzAxMjI4NDA4Mg ==&mid=2654370346&idx=1&sn=f7fa279ab7a9de50e3a59e8adc90142f&chksm= 8076e3f7b7016ae1cf479de6d8b090728a27a9bf8bef0e2d737ca8b286f08c15e- da17ce16d42 - rd.

Wang, Shiyu, and Qianhui Jin. 2013. "'Apportioning' Quotas of Mental Illness." *Southern Metropolis Daily*, October 9. http://paper.oeeee.com/nis/201310/09 /120093.html.

Yang, Jie. 2015. *Unknotting the Heart: Unemployment and Therapeutic Governance in China.* Ithaca, NY: Cornell University Press.

Zhang, Li. 2012. *In Search of Paradise: Middle-Class Living in a Chinese Metropolis.* Ithaca, NY: Cornell University Press.

Zhang, Li. 2017. "The Rise of Therapeutic Governing in Postsocialist China." *Medical Anthropology* 36(1): 6–18.

Zhang, Weixi, Shuran Li, Changhui Chen, Yueqin Huang, Jinrong Wang, Deping Wang, Jian Tu, Zuoxi Ning, Limu Fu, Liping Ji, Zhiguang Liu, Huamin Wu, Kailin Luo, Shutao Zhai, Heqin Yan, and Guorong Meng. 1998. "Epidemiological Investigation on Mental Disorders in 7 Areas of China." *Chinese Journal of Psychiatry* 31(2): 69–71.

Zhao, Shukai. 2007. "The Accountability System of Township Governments." *Chinese Sociology and Anthropology* 39(2): 64–73.

Zhou, Yingfeng. 2010. "China Will Renovate or Expand 550 Psychiatric Hospitals to Prevent Patient Violence." *China Net*, June 21. http://www.china.com.cn/policy /txt/2010-06/21/content_20305164.htm.

EMBRACING PSYCHOLOGICAL SCIENCE FOR THE "GOOD LIFE"?

Li Zhang

One summer afternoon in 2011, as I walked into a large crowded bookstore in Kunming, the provincial capital of Yunnan in southwest China where I was doing fieldwork, a book titled *Xinfu de Fangfa*, prominently figured among the bestsellers on a table near the entrance, caught my attention immediately. It was the Chinese translation of Tal Ben-Sharhar's well-known book, *Happier: Learn the Secret of Daily Joy and Lasting Fulfillment* (2007). The Chinese translation of the title is somewhat different from its original English one. It roughly means *The Methods of Happiness*. The word *fangfa* (methods, means, or ways) implies a systematic, scientific approach that can be learned and mastered through study, coaching, and practice. Therefore, the intended message to Chinese readers is that reaching the realm of happiness is not surreptitious or haphazard; instead one must rely on scientific methods. To differentiate its superior status from other numerous books on this topic, its preface is simply titled "Happiness Is a Science." But what is considered a "scientific approach" to living a happy and good life? What kind of science is reliable, and where does its power come from? What constitutes the "good life" (*meihao shenghuo*) in China today?[1] Does the promise of the "good life" bring about aspirations and hopes, or does it generate more anxiety and despair in instances where it frays? These are highly contested questions that need to be unpacked.

In the early years of the economic reform, material wealth and physical well-being weighed heavily in assessing the quality of life. Then, lifestyle choices based on housing, cars, clothing, and other consumer goods and travel experiences became important markers of a successful and desirable life (Yu 2014). But since the early 2000s and especially in the later 2010s, the vision of the "good life" has become more intricate and often involves multiple dimensions depending on one's socioeconomic background. Most noticeably, psychological well-being is gaining a salient place in the discussion on what the "good life" fully entails. Many are anxiously searching for the golden key to unlock the secret of achieving this ideal state of life. In 2016, Harvard psychiatrist Robert Waldinger, who has devoted several decades to studying what constitutes the good life, was invited to lecture in China. His status as a psychiatrist from a prestigious American university and his background in medical science give him a great deal of credibility and authority. Thus, even though his core idea—that the good life is built with good relationships—is nothing new to Chinese, it resonates well with Chinese people's thinking and is bolstered by scientific claims. It is in this context that psychological counseling and training have become new found tools for not only alleviating mental distress but also pursuing personal happiness, improving interpersonal communication skills, and enhancing family and workplace relationships among the urban middle classes who aspire for a better life beyond material consumption. This turn to one's inner landscape has given rise to a new industry catered to middle-class demands for emotional care and psy service.[2]

In this chapter, I explore two sets of questions. First, how is the notion of "science" or "the scientific" invoked by Chinese psychological experts and practitioners in their efforts to translate, brand, and apply certain branches of psychology and psychotherapy to Chinese society? I show that the popular pursuit of well-being and the "good life" in contemporary China is inseparable from the claims of modern science and Western biomedicine. Based on my long-term ethnographic fieldwork research among Chinese psychologists, counselors, psychiatrists, and urban residents in Kunming from 2010 to the present, I offer an in-depth account of how and why the so-called "science of happiness" (primarily based on positive psychology and the Satir family therapy) is surging in Chinese cities and how it is embraced by different social actors. This wave occurs under the banner of "psychological science" that some experts claim is able to effectively ease personal and social suffering. More specifically, my analysis focuses on how the notion of "science" (*kexue*) is deployed and contested by Chinese therapists, psy experts, counseling trainees, and clients of diverse backgrounds. I suggest that while scientism—the excessive belief in the power of scientific knowledge and techniques—animates contemporary urban "psy fever," it is also problematic when the actual therapeutic effects do not live up to the expectations of some clients and their family members. There is constant contestation over therapeutic efficacy and the value of new psychological methods. Thus, the so-called psychological science offers both promises and disappointments, the power to change and the pressure to conform.

Further, as I have shown elsewhere, the notion of science is not necessary in conflict with spirituality; on the contrary, some Zen Buddhist teachings of happiness have also appropriated the name of science to enhance their appeal to the broader public (Zhang 2015). The possibility of blending psychological science and spiritual practices is especially appealing for urban middle-class members who aspire for the "good life" after socialism.

The second question this chapter addresses is "how does the new wave of pursuing happiness through psychology and counseling relate to the reconfiguration of normative life conduct and the regime of self-improvement?" As I will show, the new psy experts play a key part in redefining what constitutes a normal life and a desirable emotional state of being that is good for the individual and the family. Thus, what is at stake concerns the remaking of normative subjects and life through personal acts of choice and adoption of various techniques. While these efforts are often framed by the universalizing discourse of psychology, they inevitably encounter the challenge of addressing specific social, cultural, and historical realities (see Farquhar 2001; Zhang 2014). The local articulations of what constitutes a good life and a good family are what interest me most.

In *Powers of Freedom*, Nikolas Rose suggests that therapeutics constitute a set of vital technologies for governing the autonomous self in advanced neoliberal societies. He writes, "Psychotherapeutics elaborates an ethics for which the way to happiness, or at least the conquest of unhappiness, can be specified in terms of apparently rational knowledge of subjectivity and where life conduct is to be shaped according to procedures that have a rational justification in terms of psychological norms of health and contentment" (1999, 93). In a similar vein, I argue that it is through the idiom of emotional care and self-care that Chinese counselors and therapists can acquire and exercise considerable expert authority by guiding ordinary people in overcoming difficulties, gaining happiness, and realizing their potential. In other words, such experts hold the power to construe the important question of "how to live" (Foucault 1988) and define what kind of life is normal, ethical, and worth living. Here I would like to borrow Collier and Lakoff's (2005) concept of the "regime of living" to demonstrate the profound implication of psychotherapeutic intervention in Chinese life. This intervention is much more than a medical one; rather, it offers an important means for "organizing, reasoning about, and living 'ethnically'—that is, with respect to a specific understanding of the good" (Collier and Lakoff 2005, 31). However, the regime of living is a highly contested one, and thus any significant shifts in it tend to engender struggles, negotiations, anxiety, and confusion.

Further, I must point out that the notions of individual autonomy and freedom continue to be problematic in reconceptualizing the self and the social in Chinese society. Such notions simply do not occupy the same level of salience in

China as in the Western neoliberal societies described by Nikolas Rose. In my view, what matters most today is to better understand how Chinese people inhabit and negotiate the juncture of personal autonomy and social embeddedness, self-sufficiency and familial obligations. Although the people I encountered during fieldwork are eager to learn scientific techniques to enhance their quality of life, navigating through such intricate spaces is not easy. The challenges and tensions that many people feel have largely contributed to the proliferation of psy professionals and service in Chinese cities.

Emotion-Management and "Psychological Science"

What happiness means varies greatly from person to person, but in general it refers to a satisfied state of emotion and a sense of fulfillment and well-being. Thus, it is closely related to how one manages one's emotions and psyche. When I grew up in Southwest China during the 1960s through 1980s, the idea of self-care with regard to one's emotions and feelings was rarely discussed through the lens of biomedicine in private or public life. Western-style psychological counseling was largely absent and unheard of; instead, talking to family members, close friends, and traditional Chinese medicine doctors provided some relief of mental distress. Under socialism, the concept of "mental illness" was heavily stigmatized. Referring only to severe cases of psychosis, it was commonly known as "schizophrenia" (*jingshen fenlie zheng*). In everyday life, various forms of emotional disorder were lumped into a somatized diagnostic category known as "neurasthenia" (*shenjing shuairuo*), which was to be treated by proper rest, physical exercise, and traditional Chinese medicine (Kleinman 1982; Lee 1999; Liu 1989). "Melancholia" (not "depression" which was still an unknown category) was regarded as an innate personality trait that could be affected by one's trying social and political circumstances. But there was no reference at all to any possible neurochemical unbalance or psychological dysfunctions.

Postsocialist transformations have generated heightened forms of mental distress due to increasing socioeconomic competition and pressure, but at the same time also opened up new spaces for talking about depression, anxiety, and other emotional problems, as well as about diagnosis, treatment, and coping strategies. There has emerged a robust popular desire for improving one's emotional well-being, which is crystalized in the notion of seeking personal happiness and joy among the rich and middle classes. The reform party-state has recognized the necessity and the benefit of transforming underprivileged people into "happy" (at least not disgruntled) subjects in order to maintain social stability and its political legitimacy.[3] In some cities, local government has promoted psychological self-help and sent trained counseling experts to work with the poor and the unemployed (Yang 2013). Thus, even though their motivations are quite different, these two trends—one initiated by individuals and the other by the government—are converging at the juncture of what I call "governing through happiness." Further, what connects these two efforts is the prominent place of psychological care and its claim to scientific methods in managing people's affect and reducing suffering.

The power of "science" and the idea of "saving the nation-state through science" (*kexue jiuguo*) have been around in China for over a century. At the dawn of the twentieth century, during the famous "New Culture Movement," progressive Chinese youth and intellectuals openly embraced what Chen Duxiu advocated as the saviors of the nation—"Mr. Science" and "Mr. Democracy." Their aim was to use modern Western science to create a new culture and an open, egalitarian, and liberating society, doing away with the feudalism and Confucianism that they believed were holding Chinese nation down. The trope of science stood for modern scientific principles, positivism, biomedicine, and most of all enlightenment In the midst of this grand wave, Freudian theory entered China not so much in medical and clinical practices, but "as a possible new science of the mind and a way of thinking about the characteristics of the modern person" (Larson 2009, 31). According to Wendy Larson (2009), the publication of the Chinese translation of Joseph Haven's first volume of *Mental Philosophy* as *Xinling Xue* ("The Study of the Soul/Heart") by Yan Yongjing, a missionary student who studied in the United States in the late nineteenth century, was regarded by many as the birth of psychology as a science in China (2009, 35–36). However, Freudian psychology did not take a strong hold beyond a small circle of intellectuals; its biggest influence remained within literature and literary criticism. In the early to mid 1950s, after the Chinese Communist Party's takeover of the mainland, psychologists and medical scientists largely turned to the Soviet approach, which emphasized mainly the Pavlovian behavior model. They emphasized the physiological basis for mental processes and engaged in experiments on behavioral modifications through repeated stimulus (Gao 2015). Soon after, China entered the turbulent period of the Cultural Revolution, during which nearly all Western psychologies, including the short-lived Pavlovian theory, were denounced as bourgeois false ideologies, while mental health was deemed irrelevant to class struggles and the Communist revolutionary causes. Even talk of "science" beyond Marxism and Leninism was a dangerous act and could invite harsh political criticism in those years.

The notions of "science" and "technology" regained a prominent place in the national effort to revive China when Deng Xiaoping came to power during the reform and opening era. The tenet of "Four Modernizations" advocated by the reform party-state was to lift China out of the ruins of the Cultural Revolution by stimulating economic development and engaging global modernity. This project rested squarely on full engagement of science and technology imported mainly from Euro-American countries and Japan. The main focus was on scientific innovations and technological applications that could improve infrastructures, agriculture, industries, manufactures, aerospace, education, and other material conditions of life. Most investments by the government were made in the areas of hard and applied sciences, but the application of science also began to extend to human and familial improvement as in the so-called population science (see Bakken 2000; Greenhalgh 2008; Greenhalgh and Winckler 2005).

In official discourses that permeated Chinese society, the nation's future would unquestionably depend on the power of science and technology. I still remember vividly that when I decided to take the humanities/social sciences track in high school for college preparation in the 1980s, my teachers, classmates, and other friends were puzzled and dismayed by my "unwise" choice. At that time only those with poor grades and no possibility of a bright future would go into this track, but I was ranked among the top 2 percent in my grade. My chemistry teacher, who had high hopes for me, scolded me for wasting my talent. Among some 350 students that year, only about fifty (mostly poorly performing students) went into the humanities/social sciences class. Fortunately, my parents—a professor and a high school teacher trained in Chinese literature—firmly supported my "unpopular" choice. I ended up attending China's top university—Peking University—to study Chinese literature. Only then did I receive some recognition.

As the economic reforms deepened and the material well-being of the middle classes improved considerably, there has been an increasing popular desire to use scientific knowledge and methods not only to advance material conditions and the external environment but also to enhance the inner life, which remains relatively intangible. What I have termed "the inner revolution" (*neixin de geming*) is an attempt to address mental and psychological health and subjective feelings by putting them to the forefront of the discussions of personal and national well-being (Zhang 2014, 2015). In this emerging realm, psychology is often specifically called "psychological science" (*xinli kexue*) in order to accentuate its validity and credibility. I have also observed two simultaneous processes within this inner revolution: the psychologization of social and economic issues, and the demedicalization of emotional distress (see also Kitanaka 2012). Rather than concentrating on treating mental illness, the primary drive of this movement is the promise of attaining well-being, resilience, optimism, and happiness with the aid of professionals or scientific methods of self-help.

But what constitute reliable "scientific" methods? Who gets to claim that their psychological approach is validated by science? How does one deploy scientific claims in dealing with affects and emotional care? I will explore these questions through several ethnographic encounters and argue that there is a great deal of scientism in the making—that is, the notion of "science" becomes a floating signifier that can be appropriated for different purposes without the support of any robust evidence. What is at stake is thus the ability to brand a therapeutic approach with a scientific aura and take advantage of the popular faith in science. Of course, not everyone wants to embrace psychological science as a panacea for their problems, and some have doubts or are reluctant to try it due to their different social and financial backgrounds.

Fieldwork Illumination

I did not go to Kunming to study happiness or how people try to attain the dream of the "good life." When I wrote my National Science Foundation grant proposal (which was awarded for a three-year period from 2010 to 2013), I was concerned primarily with the reported rise of depression, anxiety, stress, and other psychological problems in the postsocialist era. For someone like me who grew up in the socialist period, these were novel terms and problems in the Chinese context. I was hoping to understand how Chinese therapists tackled these problems, what kinds of therapeutic relationships emerged in the clinical setting, and how ordinary distressed people interpreted their suffering. But soon after I began my fieldwork in the summer of 2009, I discovered that a significant part of the new "psych fever" took place outside hospitals and clinics (see also Huang 2017). And what motivated many people to embrace the field of psychology and counseling was the desire to enhance their familial and social relationships, engage in the care of themselves, and improve the quality of their life. Over the past six years (mostly during the summers), although I have done some observation in the therapy clinic, I have also significantly expanded my inquiry to other arenas: professional training centers, families, corporations, police forces, schools, and so on.

In the initial phase of my research, China was already in the midst of a "psy boom," which was sweeping the cities. Bookstores carved out a new section dedicated for books and magazines on mental health, psychology, and counseling practice. Discourses on self-help, the science of happiness, and mental hygiene first targeted middle-class urbanites and then trickled down to marginalized social groups through local government programs (see Yang 2015). I saw many newspapers and some taxicabs carrying advertisements for private counseling services, training workshops, and websites on psychological well-being. International psychological experts on how to raise a smart and successful child and how to create a happier life often attracted large crowds. More and more people were willing to pay a hefty ticket price (ranging from 200 to 800 yuan) for these workshops on self-development, life coaching, and parenting.

It has become clear to me that the counseling movement is not just a medical/clinical phenomenon but a core component of the broader "inner revolution" that is reconfiguring selfhood, social life, and politics across Chinese cities. Moreover, this time with the aid of experts and the promise of *xinli kexue* (in particular, positive psychology and the Satir model of therapy), the expectation for quick, positive changes in one's affect and social life is higher than ever before.[4] It is in this context that I seek to explore the slippery relationship between science and the pursuit of happiness through a demedicalized form of counseling I observed in my ethnographic encounters.

Anxious Mothers, Stressed Kids

With their eyes closed, twenty women and three men, aged from the mid-twenties to early forties, were standing in a circle to practice guided relaxation meditation led by the soft voice of Ms. Chen, who was the instructor of a workshop titled "Accompanying Children as They Grow Up." I was among them in the circle following the instruction. I had known Chen for two years, since the beginning of my preliminary research. She was a middle-aged lecturer at a local college but devoted a great deal of her time and energy to psychological counseling and training. Certified ten years ago after attending an intensive three-month preparation program, she joined a counseling team that secured the right to establish the Satir Center of Southwest China. The Satir family therapy model, which was developed by an American psychotherapist, Virginia Satir (1916–1988), is one of the most popular approaches adopted by therapists in Kunming and many other cities (see Zhang 2014). With its explicit emphasis on a systematic and experiential approach to therapy, this model has been adapted and widely used in different contexts such as schools, firms, police, military, and other individual-based workshops to improve communication, parenting, education, and personal growth. While the Satir model was popular in the United States from the 1960s to 1980s, today few people know about it. But in China, it is spreading like wildfire partly due to its special attention on family dynamics and parenting skills.

After the short meditation period, the participants were asked to turn to the nearby person to introduce themselves and share their life experiences and parental styles. Most of these participants were mothers who were having a hard time communicating with and relating to their children. They were anxious to learn how to improve their parental skills in order to help their children succeed. The woman next to me, Ms. Jiang, told me that her son was admitted to the best high school in Kunming because of his special athletic skills, but he struggled hard and could not keep up with the basic academic mandate. "Under the pressure, his personality has changed so much. I used to know my son very well—what he was thinking and what he liked, but now he will not tell me anything. I feel very sad about his withdrawal but do not know how to help him," said Jiang with tears rolling down on her face. She told me that her motivation to attend this workshop was to learn some credible methods from the experts so as to cultivate a healthier relationship with her son and help ensure his academic success. Jiang is certainly not alone in facing such a dilemma. China's "examination regime" (especially the famous *gaokao*, the national college entrance exam) gives utmost importance to test scores, which largely determine one's career opportunity and future life (Kuan 2015; Fong 2004). Parents are often torn between the impetus to push their kids to work harder in order to attain higher scores and class ranking and the desire to preserve their beloved kids' "psychological health" and happiness. Jiang expressed this deep tension between love and discipline that tormented her this way:

> I have lately realized that whenever I speak to my son, I am always urging him to study harder, checking his homework and test results, and warning him not to waste time on nonschool related activities. I rarely talk to him about other things or show that I care about his feelings, interests, and social life. I fear I might be too rational and too practical, and not be loving and warm enough. But what can I do as a mother living in this highly competitive, test-score-driven society? I am here hoping to learn some techniques from experts to change this situation and improve our life quality.

It was not hard to detect a profound sense of guilt, frustration, and helplessness in her voice. During the course of my fieldwork, I encountered many mothers like Jiang who sought psychological counseling or training as their last resort in the hope of improving their troubled relationship with their kids or "saving" their kids from what they perceived to be a failing path. Often, the yard stick used to measure a student's future success and worth was test scores. Those with low test scores are usually labeled as "problematic" or "bad," and thus not deserving full parental love. They are scolded constantly by their parents and teachers, while carrying a great burden of shame. Parents are deeply conflicted (*maodun*) in their hearts and are not sure what are the effective ways of transforming their children. They talk about their painful struggle between the divergent directions in which their hearts and minds want to take them: a "tiger mom" for success or a loving mom for well-being.

At the workshop, participants also learned the four main goals of healthy personal growth advocated by the Satir teaching: to enhance self-worth, to gain more choices, to become more responsible, and to reach greater harmony (personal, familial, and social). Chen told the anxious mothers there: "Your children's problems are not problems; what is problematic is the way you react to them. You must remember that changes are always possible. It is true that we cannot alter what happened in the past, but we can certainly change how they impact us today." She then assured them that the Satir family therapy, which was tested over time and validated by scientific principles in the United States and beyond, could be applicable to a variety of situations far beyond the clinical setting. She continued, "The Satir methods can help you untie the knot of your child's heart, and cultivate a sense of self-worth and responsibility in them."

Chen was passionate and empathetic during the two-hour workshop. She took two cases from the participants and asked them to share their dilemmas and engaged them openly in an interactive way. For example, one of the woman said emotionally: "One day my child burst out that he hated me and wanted me out of his life. When I heard those words, I was devastated. He has become a withdrawn and angry teenager. I asked myself when I lost the child I had raised and how I could get the child I used to know back." Chen asked her to recall when was last time she had a relaxed conversation with her son about things he liked, or they enjoyed something together. She said that she did not remember any and that their life revolved mostly around meeting his basic needs (food and clothes) and ensuring his school work was done on time. "These days everything I say seems to be wrong or boring to him. He is rebellious and just wants to hang out with his friends and listen to pop music. His grades are falling, and I cannot do much about it," said the sobbing mother.

After attentively listening for a while, Chen started to coach this woman how to craft a communication channel with her son without rushing into the mode of lecturing and nagging. They practiced role-playing by using a Satir technique called "family sculpturing." This mother was asked to play the role of her son while Chen acted as the mother. The goal was to let her feel for a moment what it was like to inhabit the son's position facing school pressure and an anxious and demanding mother. In the process, the woman started to weep: "I knew my son was under a lot of pressure, but I did not realize what it felt like. I can be more compassionate rather than just adding more stress to him,"

At the end of the workshop, Chen recommended to the group a couple of Satir's books already available in Chinese translation. She reminded them that the aim of the workshop was "about how to *accompany* your child (*pei haizi zhangda*) to flourish and develop, not about managing, lecturing, or controlling them." This statement may sound banal in the United States, but it represents a radical shift in the parenting mode in China. This new style of parenting through a relatively egalitarian, open, and nurturing relationship is quite different from a traditional one, which is largely hierarchical, disciplinary, and grades-focused. It also redefines "success" in terms of raising a competent, happy, and emotionally balanced child rather than solely in terms of school performance. One young mother whose child was just one year old told me that this was exactly what she wanted to learn—the right way of raising a healthy child and grow with her child before it was too late. She did not want to treat her child the way she was treated by her authoritarian parents.

Counseling Family Harmony

On our way to a house-call counseling session, Mr. Ling gave me a quick briefing on his client and the family situation. The woman we were about to see, Mrs. Xiong, retired, was in her late seventies. She and her husband had their own home in Sichuan Province but spent several months with each of their three children's families every year. They were currently staying at the house of the youngest daughter, who was an executive assistant to the president of a local university. Mrs. Xiong was reported to have a terrible temper, highly critical of everyone around her, needy, and so overly controlling of her husband that he could not leave the house alone most of the time. Her children hired three domestic workers to help take care of their aging parents, but she drove each one away due to her temper and verbal abuse. The two daughters were afraid that their mother would eventually suffocate their father, who already suffered from other illnesses, and drive the current helper away. Since they had a difficult time interacting with their angry mother, they could not enjoy life. Running out of ideas and patience, as a last resort, they hired Ling to help because of his reputation in effective family therapy.

I have known Ling for several years and helped his small firm with some translation work in his effort to acquire a franchise for establishing a Satir Model Training Center in Southwest China. The effort was successful, and he became a popular guru of teaching the Satir therapy model in the province. He is sharp, pragmatic, and passionate about spreading the Satir model. Prior to becoming a psychotherapist, he was a college lecturer in the political thought work division. He taught courses on Marxism and Chinese Communist Party (CCP) ideologies, which were required for all college students in those years. A big draw to psychology for him was its scientific basis and pragmatic application, but he also acknowledged that his previous job in political thought work prepared him well for his current counseling work. "My earlier work made me a good listener and better at connecting with all kinds of people, but political thought work itself was very dry and boring to students and myself. More importantly, the big difference is that thought work is about ideology while psychological counseling is based on science," claimed Lin with confidence. In our conversation, Ling often forgot or intentionally bypassed my suggestion that scientism itself was also a form of ideology. It was clear that for him (and most Chinese therapists) "ideology" is closely tied to political thought and thus bad or uninviting; "science," on the other hand, is regarded as intrinsically good. In any case, he was passionate for doing counseling and found it gratifying as he often saw positive changes in the clients and gained a great deal of respect from them.

Indeed, "political thought work" is an invention created by the CCP based on the vision of the explicit task of creating the "collective," which requires the shaping and reinforcing of the correct individual mindset so as to ensure a successful revolutionary project. By contrast, "scientific counseling" is more in line with the ethos of capitalist modernity that underlines personal contentedness and fulfillment. But political thought work shares certain similar features with cognitive behavioral therapy in that both insist that thought patterns can be altered and such modifications can lead to behavioral change.[5]

As we entered the client's house, the two daughters greeted us and brought their parents out of the bedroom. The mother was short, skinny, and seemingly distraught. She treated Ling like a medical doctor and began to talk about her symptoms: "I am suffering a lot these days. I have insomnia and have to rely on medications. I have high-blood pressure and have trouble urinating sometimes." Since she worried about her health, she wanted to test her blood pressure repeatedly every day. She appeared anxious and irritable and said that her left hand was numb and rigid. I also noticed a middle-aged woman sitting by the side who was later introduced as the helper. She reported that whenever Xiong was nervous, she would have trouble urinating. "I try to encourage her that she can do it if she just relaxes a bit. But she gets mad at me and quarrels with me, accusing me of not taking good care of her and telling me to go home." As the woman spoke, Xiong started to point a finger at her saying she was a bad person and a liar. She also accused her daughters of lacking care and not standing up for her.

Ling spent the first thirty minutes listening to each of them talking about their concerns. Based on that, he felt that the crux of the problem was a pathological family dynamic pattern coupled by poor communication skills. For instance, they rarely looked at each other directly when they talked. Their communication was usually through arguing. They could not see the other's point of view but focused instead on their own misery. In addition, Ling felt that Xiong clearly suffered from anxiety but tended to somatize her distress through physical discomforts and use these them to demand constant attention and care.

Later I learned that Xiong suffered a lot during the Cultural Revolution due to her husband's unfavorable political background and the pressure of raising three children in a shortage economy. She became moody, harsh, and distrustful of others. As she grew older, she felt that it was her children's time to pay back—to show their filial piety and take care of their aging parents. Thus, she refused several domestic workers her children hired to take care of her and her husband; instead she demanded that the children, not some hired strangers, performed their duty. Once she made a suicide threat claiming that she would overdose herself by taking *anding*, a common tranquilizer/sleep aid. She kept saying, "Old people are pitiful. Children nowadays are no longer filial. It is better to die than to live."

Ling started by teaching Xiong basic relaxation techniques through breathing and visualization to help her solve some practical problems such as urination difficulty. She was a bit reluctant at first but went along because her daughter was urging her to give it a try: "Teacher Ling's method is scientifically based. It will work!" Then she pointed at me saying, "Teacher Zhang is a researcher from America. Ask her if you do not believe."[6] I felt a bit awkward about the claim but nodded, "Well, it certainly worked for my mother." Then, Ling turned to work on their communication. He asked Xiong and the woman helper to look into each other's eyes directly and state their wishes calmly without yelling and accusation. It was not easy and took a while for them to be able to engage each other in this way. The helper said that she wished that Mrs. Xiong would not lose temper at her easily and be more patient. Xiong said that she wanted the woman to follow her will better and be kind and honest. But quickly she became emotional and started to point her finger at the woman again. Ling had to remind her several times of the agreed protocol for dialog. He asked her how she felt as he played the role of the worker and pointed his finger at her. She said that made her feel hurt and she had not realized how this gesture would hurt others deeply.

Next it was the daughters' turn. "Tell your mother what changes in the family you really want to see and why," said Ling. The youngest daughter kneeled down in front of her mother and said in tears: "What I want is to see you and Dad living a comfortable and normal good life. Materially we do not lack anything, but emotionally we are tormented by the tensions in the family and your anger. Please accept Teacher Ling's expert help and teaching. Let us work together to improve our quality of life." Xiong was somewhat touched, nodded, and promised to work with others. She was willing to accept the coaching help because Ling appeared to be fair, compassionate, and skillful. Throughout the two-hour session, her husband, a small old man, was quiet most of the time and appeared intimidated by her. Only towards the end he chimed in: "Xiao Wang [the helper] is actually pretty good. We should treat her well and respect each other. On your part, Xiao Wang, you could be more active, tolerating, and diligent. Just take good care of her; don't worry about me."

Xiong never smiled during the entire time, but her way of communication shifted from accusing and cursing to a somewhat calmer mode of talking about the situation and her feelings. That was an important first step. Ling also coached the family on understanding, acceptance, and potentiality in the spirit of positive psychology and the Satir model. They settled on a set of basic rules of communication in the family. At the end, everyone agreed to have Ling come back to do more coaching work over the next two months. They acknowledged that positive changes come from a joint effort by everyone, not just one individual, even though Xiong was initially the primary client in need of help for this session.

On my way back home, I could not help but think about a similar yet also quite different situation that occurred some forty years ago. My father was a department chair at Yunnan University for a long period. In those days, a chair was expected to help faculty and staff in mediating family disputes, marital problems, and other interpersonal troubles. I remember clearly the many hours my father spent speaking with distraught faculty or their spouses in our tiny apartment, where I had no space to hide. Essentially people like my father played the role of a mediator and a counselor without having any formal training in psychology. Today it is very rare for a department chair or other local cadres to perform such functions. Deeply troubled families tend to use self-help techniques or go to professionals like Ling to deal with their problems, which are at once social, psychological, medical, and personal. There is also a common belief today that experts who are perceived as equipped with some special tools can fix their problems more effectively and protect their privacy even though the costs are high. Creating and maintaining family harmony is emerging as an indispensable part of the "good life." But only the rich and the middle class can afford to obtain professional service and technical tools; this option is beyond the reach of most poor and working-class families.

Coaching Resilience

Over the past two decades, the notion of "trauma" (*chuangshang*) has been widely recognized as a legitimate medical and psychological problem among soldiers, police officers, and firefighters, who are regularly sent on disaster relief tasks in China. Witnessing earthquake devastation, horrific traffic accidents, and fire incidents and engaging in life-threatening rescue work often lead to what is now known as post-traumatic stress disorder (PTSD). For a long time, such troubling psychological impact was not recognized or talked about as rescuers were sup-posed to be tough and unshakable in the public and official eyes. With the spread of psychological counseling in the cities, government officials have gradually re-alized the need to pay attention to the effects of trauma and ordered relevant organizations to incorporate trauma prevention and healing into their training. This call for utilizing psychological science in this context is unprecedented in China.

Mr. Yang is a therapist in his forties and the head of a local counseling train-ing center in Kunming. I have known him for six years and participated in nu-merous counseling training workshops he offered. Since 2011, he has been con-sulted by several fire departments in the city and invited to hold workshops and conduct counseling. He explained to me why such needs are growing in this profession:

> Nowadays firefighters' tasks include not only combating fires but also disaster-rescuing, especially since the massive Wenchuan earthquake. When they encounter very dangerous or gruesome scenes of death and injuries, it is hard not to be impacted deeply. Sometimes, they see their own friends tragically dying during the rescue too. This is particularly difficult for younger fighters who are not even twenty years old. They are not prepared to handle such harsh reality and are so traumatized that they cannot eat or sleep for several days afterward.

When Yang first started this line of work, he did it on the volunteer basis out of his good heart. He was sympathetic toward the young men who risked their lives and suffered from distress. But soon he realized that he needed two assistants and money to cover the cost and his time. So he began to charge a fee, although it was not much compared with what he would charge a private firm. Most fire departments want him to offer training twice a year or after a major disaster-relief event. His teaching focuses on how to recognize PTSD symptoms, how to prevent and treat them, and stress management. "There was simply no psycho-logical coaching for these people. Political thought work was supposed to take care of all problems before. There was no science involved. When some of them experienced PTSD or a panic attack for the first time, they were frightened and did not know what was wrong with them," said Yang. "Now after attending my workshops, many of them want to learn more psychology because they see the benefits and hope to master some useful tools." One day he was surprised to hear the team leader saying, "Teacher Yang, we have lately realized that we all have psychological problems to a varying degree, and no one should feel shamed about it."

During his workshops, Yang did not just lecture, but also relied on frequent interactions with individual participants to demonstrate how he would coach social and emotional issues. This participatory approach was welcomed by the fire fighters because they were not treated as passive recipients, but as active collaborators. His choice of words in his training was also important. To differentiate his training from past political ideology propaganda work, he used *celiang* (measure), *pinggu* (evaluate), *fangfa* (method), and *tiyan* (experience)—terms recognized by Chinese people as technical and methodical—to highlight the scientific nature of his approach. Of course, there were limits to this approach since some firefighters were very concerned about revealing their true feelings and secretes in front of others. But he was able to make them feel comfortable and gradually open up to a certain degree. Such workshops provided a relatively safe space for them to explore psychological issues and familial problems, which was not available before. He was trying to strike a good balance between sharing private feelings and protecting one's privacy. He used a hybrid form of counseling methods ranging from talk therapy, sand-play therapy, and art and music therapy, all of which are relatively new yet appealing to Chinese clients.

Local fire departments, along with some other work units, have begun to conduct psychological evaluations among their employees to identify potential problems and offer resilience training (Zhang 2015). This form of governing through psychological science is new to China and remains experimental. According to Yang, most firefighters he met appreciated the opportunity to gain new psychological knowledge and acquire tools to better understand their emotional state and reactions to trauma. But some remained skeptical that such psychologically oriented workshops could actually alleviate their suffering and believed that resilience was an innate quality one was born with or not. If they do not experience positive changes in themselves after training, all the scientific claims made by the therapists become meaningless. They want to see real results in their life! From the managers' and party officials' point of view, creating a more knowable and governable workforce is desirable only if it does not generate unwanted side-effects (such as more reflexive and defiant employees) that turn out to be more challenging to manage.

Conclusion

For centuries countless people all over the world have tried hard to attain and maximize their well-being and happiness. Chinese people are no exception. Yet, as I have briefly outlined, there has been a significant shift starting in the 2000s from prioritizing collective well-being and national welfare to focusing on individual happiness and self-growth defined among Chinese by a sense of contentment, fulfillment, and family harmony. And this shift has been mediated by the emergence of scientific discourses on mental health and new therapeutics. As early as the May Fourth Movement at the turn of the twentieth century, the notion of "science" was invoked as a critical path to save the Chinese nation. Later, under Maoist high socialism, science was relegated to a bourgeois realm or replaced with Marxist ideologies. Since the economic reform initiated by the Deng Xiaoping regime, science and technology have regained a salient place in rebuilding a stronger China and maintained an authoritative hold on discourses on national rejuvenation.

Today, what is most fascinating in my view is the application of human and health sciences to people's interior lives—their hearts and minds and the extent to which psychological science in particular is seen as vital for individuals and families to flourish and achieve the "good life." This vision of governing through science and the merging of national and individual well-being is celebrated by President Xi Jingping in his call for a new China dream. I suggest that the effectiveness of this emerging biopolitical governance in the name of care and science remains unclear and deserves more of our research attention (see Borovoy and Zhang 2017).

Since science is always contingent and contextual, a gap between its claims and effects is not unusual. But in the domain of popular psychology and counseling, I find Chinese practitioners' relationship with science even more tenuous for several reasons. Many of them lack solid training after only several months of crash courses and then go into practice with insufficient skills and knowledge. The scientific quality of their work is frequently asserted, but is not adequately substantiated by concrete evidence or therapeutic efficacy. This is a main concern for many potential Chinese clients, yet curiously they do not seem to be too interested in questioning scientific claims per se. Instead they are more concerned with the therapist's reputation and status. Further, psychological processes and emotional states are highly subjective and sometimes erratic experiences that are hard to be measured and evaluated. Given the short history of psychology and psychotherapy in China, practitioners have not developed a robust means to document and demonstrate treatment efficiency in a systematic manner. Thus, because so far therapeutic efficacy is very much individually based, if counseling does not live up to its promise in some cases, the client is often told that the problem resides in his or her effort, not in counseling itself. These circumstances leave plenty room for the practitioners to make promises and over-sell the value of "psychological science" to the middle class's pursuit of happiness and the "good life."

NOTES

An earlier version of this chapter was presented at the conference on "A Better Life through Science and Biomedicine?" held at Harvard University, April 15–16, 2016. I thank all the participants, especially Susan Greenhalgh, for their engaging and helpful comments. I am also grateful for the two anonymous reviewers for the press who provided insightful suggestions that I incorporated into the final version of this chapter.

1. Anthropologist Arthur Kleinman has suggested that "What is an adequate life in China?" is a central question to explore (see Zhang, Kleinman, and Tu 2011). While I agree with his view, the notion of the "good life," not just an adequate life, has been gaining public attention in China and reshaping the imagination of the urban middle-classes. Lauren Berlant's (2011) notion of "cruel optimism" is quite relevant to our attention to the "good life," yet it has its limit in the Asian context (see Borovoy and Zhang 2017).

2. This shift is part of a larger trend of what Farquhar and Zhang (2012) have termed "nurturing life"—a Chinese way of self-help or self-health. What distinguishes the cases I show in this chapter is how deeply psy experts mediate this self-practice.

3. This new focus on personal happiness is in contrast with another kind of affect—revolutionary fervor incited by high socialism (Dutton 2005).

4. Martin Seligman is widely recognized in China as the founder of positive psychology. All his major works such as *Learned Optimism* (1991) and *Authentic Happiness* (2002) have been translated into Chinese and sold well in China.

5. See my article (Zhang 2014) that examines the close relationship between socialist thought work and certain strands of psychotherapy in China.

6. In China, it is common for people to refer to psychotherapists and counselors as "teachers" to show respect and give them an aura of authority.

References

Bakken, Børge. 2000. *The Exemplary Society: Human Improvement, Social Control, and the Dangers of Modernity in China.* New York: Oxford University Press.

Ben-Sharhar, Tal. 2007. *Happier: Learn the Secret of Daily Joy and Lasting Fulfillment.* McGraw Hill Education.

Berlant, Lauren. 2011. *Cruel Optimism.* Durham: Duke University Press.

Borovoy, Amy, and Li Zhang. 2017. "Between Biopolitical Governance and Care: Rethinking Health, Selfhood, and Social Welfare in East Asia." *Medical Anthropology* 36 (1): 1–5.

Collier, Stephen, and Andrew Lakoff. 2005. "On Regimes of Living." In *Global Assemblages: Technology, Politics, and Ethics as Anthropological Problems*, edited by Aihwa Ong and Stephen J. Collier, Malden, 22–39. Malden, MA: Blackwell Publishing.

Dutton, Michael. 2005. *Policing Chinese Politics.* Durham: Duke University Press.

Farquhar, Judith. 2001. "For Your Reading Pleasure: Self-Health (*Ziwo Baojian*) Information in 1990s Beijing." *Positions* 9 (1): 105–130.

Farquhar, Judith, and Qicheng Zhang. 2012. *Ten Thousand Things: Nurturing Life in Contemporary Beijing.* New York: Zone Books.

Fong, Vanessa. 2004. *Only Hope: Coming of Age under China's One-Child Policy.* Stanford: Stanford University Press.

Foucault, Michel. 1988. "Technologies of the Self." In *Technologies of the Self: A Seminar with Michel Foucault*, edited by L. H. Martin, 16–49. London: Tavistock.

Gao, Zhipeng. 2015. "Pavlovianism in China: Politics and Differentiation Across Scientific Discipline in the Maoist Era." *History of Science* 53 (1): 57–85.

Greenhalgh, Susan. 2008. *Just One Child: Science and Policy in Deng's China*. Berkeley: University of California Press.

Greenhalgh, Susan, and Edwin Winckler. 2005. *Governing China's Population: From Leninist to Neoliberal Biopolitics*. Stanford: Stanford University Press.

Huang, Hsuan-Ying. 2017. "Therapy Made Easy: E-Commerce and Infrastructure in China's Psycho-Boom." *China Perspectives* 4: 29–39.

Kitanaka, Junko. 2012. *Depression in Japan: Psychiatric Cures for a Society in Distress*. Princeton: Princeton University Press.

Kleinman, Arthur. 1982. "Neurasthenia and Depression: A Study of Somatization and Culture in China." *Culture, Medicine and Psychiatry* 6 (2): 117–190.

Kuan, Teresa. 2015. *Love's Uncertainty: The Politics and Ethics of Child Rearing in Contemporary China*. Berkeley: University of California Press.

Larson, Wendy. 2009. *From Ah Q to Lei Feng: Freud and Revolutionary Spirit in 20th Century China*. Stanford: Stanford University Press.

Lee, Sing. 1999. "Diagnosis Postponed: Shenjing Shuairuo and the Transformation of Psychiatry in Post-Mao China." *Culture, Medicine and Psychiatry* 23: 349–380.

Liu, Shixie. 1989. "Neurasthenia in China: Modern and Traditional Criteria for its Diagnosis." *Culture, Medicine and Psychiatry* 13: 163–186.

Rose, Nickolas. 1990. *Governing the Soul: The Shaping of the Private Self*. London and New York: Routledge.

Rose, Nikolas. 1999. *Powers of Freedom: Reframing Political Thought*. Cambridge, UK: Cambridge University Press.

Seligman, Martin. 1991. *Learned Optimism: How to Change Your Mind and Your Life*. New York: Knopf.

Seligman, Martin. 2002. *Authentic Happiness: Using the New Positive Psychology to Realize Your Potential for Lasting Fulfillment*. Tampa: Free Press.

Yang, Jie. 2013. "Fake Happiness": Counseling, Potentiality, and Psycho-Politics in China." *Ethos* 41 (3): 291–311.

Yang, Jie. 2015. *Unknotting the Heart: Unemployment and Therapeutic Governance in China*. Ithaca, NY: Cornell University Press.

Yu, LiAnne. 2014. *Consumption in China: How China's New Consumer ideology is Shaping the Nation*. Cambridge, UK: Polity.

Zhang, Everett, Arthur Kleinman, and Tu Weiming, eds. 2011. *Governance of Life in Chinese Moral Experience: The Quest for An Adequate Life*. London and New York: Routledge.

Zhang, Li. 2011. *In Search of Paradise: Middle-Class Living in A Chinese Metropolis*. Ithaca, NY: Cornell University Press.

Zhang, Li. 2014. "Bentuhua: Culturing Psychotherapy in Postsocialist China." *Culture, Medicine, and Psychiatry* 38 (2): 283–305.

Zhang, Li. 2016. "The Rise of Therapeutic Governing in Postsocialist China." *Medical Anthropology* 36 (1): 6–18.

NEGOTIATING EVIDENCE AND EFFICACY IN EXPERIMENTAL MEDICINE

Priscilla Song

The transition from laboratory bench to hospital bed is happening at an accelerated pace in contemporary China, with increasing numbers of new treatments being tested on patients. As the former emphasis on state-funded preventive care has yielded to a market-driven pursuit of high-tech interventions and subsequent efforts to redress growing health inequalities, ambitious Chinese clinicians and their increasingly globalized clientele have latched onto biomedical innovation as their ticket to personal success. In this chapter, I examine how urban Chinese neurosurgeons have leveraged fetal cell therapies in order to survive and thrive in a rapidly changing healthcare system. Their experimental practices raise key questions about the ethics and epistemology of clinical experimentation at the "cutting edge" of biomedical practice in contemporary China.

As the other chapters in this volume demonstrate across diverse realms, the pursuit of science and technology is not just a strategy for individual triumph but a broader cultural mandate for national salvation (which I have described elsewhere as "technonationalism" [Song 2017, 107]). While changes in the financing and organization of healthcare sparked by China's transition to market socialism over the past few decades have enabled and encouraged entrepreneurial physicians to develop advanced therapeutic interventions, these high-tech desires must be situated in a broader Chinese program of scientific and technological modernization. Social historian David Rothman (1997) has traced the technological imperative in American healthcare, tying the pursuit of advanced medical technologies to a history of American exceptionalism rooted in middle-class priorities. The Chinese passion for science and technology arises from a very different cultural sensibility: a history of geopolitical humiliation framed in terms of medical pathology. Branded the "sick man of East Asia," late imperial China suffered crushing setbacks from technologically superior Western nations. Engineering dreams of national ascent by "governing through science" (Greenhalgh, Introduction, this volume) have animated successive Chinese regimes, from the imperial to the (post)socialist.

As ambitious clinicians have sought to mobilize science and technology to remake Chinese healthcare, they have faced resistance in an increasingly globalized context. The field of biomedical research has proven particularly contentious, with Chinese physicians and scientists facing challenges from international colleagues about the integrity of their data, the reliability of their results, and the ethics of their practices. Medical entrepreneurs experimenting with new biomedical interventions are particularly susceptible to accusations of quackery as they negotiate the overlapping and at times competing motives of helping patients, advancing science, and making a profit. Researchers and regulators in the United States and Europe have lambasted Chinese practitioners for moving too quickly into human therapies without subjecting their treatments to the rigors of evidence-based medicine. Without proof in the form of placebo-controlled clinical trials, the critics charge, Chinese clinicians must be swindlers preying on the desperation of patients.

The charge of medical fraud carries particular resonance in China. While the focus on evidence-based medicine is new, charges of quackery are not. Writer and social critic Lao She satirized Chinese medical services in the 1930s in his short story "A Brilliant Beginning," which describes a group of charlatans who pose as physicians and try to make a profit by performing fraudulent injections (Lao 2011). Since the reintroduction of private enterprise in postsocialist China beginning in the 1980s, the country has been plagued by tales of counterfeit goods, from bootleg DVDs and pirated software to imitation handbags and fake cigarettes. The epidemic of fake and shoddy products produced in China has been particularly problematic in the medical realm and has affected both domestic and foreign patients around the globe. Transnational scandals have included Chinese cough medicine laced with diethylene glycol and exported abroad, which fatally poisoned more than one hundred Panamanians in 2006 (Rentz et al. 2008). Counterfeit heparin, a blood thinner produced in China for the pharmaceutical giant Baxter, was linked to nearly a thousand cases of serious injuries reported to the U.S. Food and Drug Administration (Harris and Bogdanich 2008). Even infant formula has not been exempt from the tampering of unscrupulous manufacturers seeking to increase their profit margins. The label "Made in China" has thus become synonymous with suspect imitations that are not just poor quality but have even injured or killed thousands worldwide. Regulators, trade officials, and journalists have blamed the problem on corruption, cutthroat competition, and a lack of regulatory oversight, which have encouraged manufacturers to cut corners in the race to the bottom line. Given this context of global suspicion of Chinese products, it is not surprising that experimental biomedical treatments in China have been interpreted as ploys by profiteering medical entrepreneurs. Yet thousands of patients from more than eighty countries have sought fetal cell transplantation in Beijing. Although critics have been quick to dismiss these biomedical odysseys as a matter of desperate patients duped by false hopes, the nuanced deliberations and complex experiences of the people I met in Beijing and online deserve more careful consideration.

In this chapter, I take a closer look at the semiotics by which practitioners and patients recognize medical authenticity and charlatanism. Based on two years of fieldwork conducted between 2005 and 2007 and several shorter research trips carried out through 2017, I provide ethnographic insight on how clinicians and patients at the pseudonymous New Century Neurosurgery Clinic in western Beijing have interpreted signs of evidence, negotiated standards of proof, and resolved questions of efficacy in the transnational realm of experimental medicine. The question "Does it work?" permeated my conversations with clinicians, patients, and family members involved with this experimental neurosurgery clinic. I document how New Century clinicians have contested international biomedical research protocols and devised new methods to assess the effectiveness of their experimental fetal cell transplantation surgery. These strategies included developing improved measurement scales to capture postsurgical differences in bodily function and feeling, inviting foreign doctors to witness clinical procedures for themselves, conducting follow-up studies with former patients, and attempting to publish scientific reports in internationally credible journals. As this chapter will demonstrate, their efforts to defend the integrity of their data and the ethics of their practices have been exacerbated by the problem of unequal resource distribution in rural and urban China. By examining the challenges New Century clinicians have faced in documenting the effectiveness of their fetal cell transplantation procedure, I demonstrate how new modes of validation are emerging as viable alternatives to the hegemonic discourse of randomized controlled trials that has dominated the quest for "evidence" in experimental medicine.

Evaluating Efficacy

American and European clinicians have contested the legitimacy of Chinese fetal cell therapies, characterizing them as "invasive cell implants for neurologic diseases" and an "unsubstantiated clinical practice" (Dobkin, Curt, and Guest 2006,

6, 13). Feminist scholar Charis Thompson has documented the heightened moral language surrounding these critiques, observing that the American scientific community has written off so-called stem cell tourism as an "off-shore hazard to good science" marketed by "rogue" clinicians (2013, 118). These critiques hinge on the charge that these Chinese experimental procedures have not followed "international standards for scientific trial methodologies" (Dobkin, Curt, and Guest 2006, 13). These critiques must be contextualized within a broader historical framework that addresses how guidelines governing medical research in the United States have developed.[1] Researchers, clinicians, and pharmaceutical companies seeking to market a new medical intervention in the United States must first perform laboratory and animal tests to determine how the treatment works and whether it is likely to be safe and effective in humans. In order to obtain approval to test the treatment on people, researchers must then submit an Investigational New Drug application to the Food and Drug Administration. Both the conduct and evaluation of these clinical trials are monitored to ensure the protection of human subjects and the collection of reliable data on the safety and efficacy of the treatment under question. The focus on efficacy in the context of clinical trials is significant. As deployed by clinical researchers, "efficacy" refers to the measurable effects of a specific intervention under controlled conditions (Cochrane 1972; Higgins and Green 2011). This is different from the effect that the intervention might have under ordinary circumstances in the real world, which in the world of biomedical research is described as "effectiveness" (Institute of Medicine 2009).

This is not a trivial distinction, as discussions of biomedical "efficacy" and even "effectiveness" set aside many of the real-world contexts under which a new intervention might be administered and taken. The sociopolitical is one crucial context often neglected in discussions of biomedical efficacy. This is particularly evident when we look at how other disciplines have framed efficacy. Anthropologists, for example, have explored broader questions of meaning and effectiveness in the context of social action. These approaches challenge the reductionist parameters of biomedical frameworks. Concerns about whether a medicine or healing practice "works" extend far beyond the narrow constraints of laboratory or clinical paradigms. As Sienna Craig (2012) highlights in the case of Tibetan medicine, efficacy is an intersubjective phenomenon negotiated between practitioners and patients in the context of specific "social ecologies." This expanded understanding of efficacy—as the capacity for producing a desired result or effect—takes into account the social and political contexts in which knowledge claims are made, by whom, and for what purposes.[2] The fact that clinical trial proponents have convinced U.S. regulators and the public to accept such a narrow definition of efficacy points to the ascendancy of laboratory-based methods in biomedicine today.

Described by historian Ilana Löwy (2000, 49) as a key technology in moving medicine from being an idiosyncratic "art of healing" to an exact science, randomized controlled trials (RCTs) developed over the past half-century as an objective way for doctors and drug manufacturers to evaluate proliferating drug therapies and provide proof of safety and efficacy. As the name suggests, RCTs involve the random assignment of research subjects to comparison groups. The intervention in question is compared against a standard given to a control group, generally an inactive substance known as a *placebo*, thus enabling researchers to evaluate differential effects. RCTs also provided clinicians with a convincing method for establishing their own credibility, deflecting suspicion, and distinguishing themselves from fraudulent patent medicine peddlers. Historian Harry Marks (1997) has documented how RCTs emerged through the process of "therapeutic reform" to become the key principle in experimental medicine. Broader shifts in the organization of medical practice and research in the United States contributed to entrenchment of RCTs—in particular the intensified involvement of the American government in biomedical research, the organization of healthcare after World War II, and the growth of academic medicine (Marks 1997).

The rise of the evidence-based medicine (EBM) movement in the 1990s to make healthcare a more rational and cost-effective practice (Guyatt et al. 1992) has further solidified the status of RCTs in clinical research and practice. Proponents have described EBM as a "paradigm shift" that "de-emphasizes intuition, unsystematic clinical experience, and pathophysiologic rationale as sufficient grounds for clinical decision making and stresses the examination of evidence from clinical research" (EBM Working Group 1992, 2420). Medical historians have challenged this characterization of EBM as a groundbreaking phenomenon, noting a history of experimentation in medicine that emphasized the importance of compiling clinical data based on comparative observations since at least the eighteenth century in Europe (Bynum 2006; Tröhler 2005). Regardless of historical accuracy, proclamations about EBM as a scientific revolution in medicine have played a key role in elevating RCTs to the "gold standard" for determining the efficacy of new treatments.

Contesting Sham Surgery

The use of randomized, controlled clinical trials to test the efficacy of new treatments originated from a particular context: American pharmaceutical industry regulations during the twentieth century. But drugs are qualitatively different from surgical procedures. Surgeons historically have utilized very different methods for testing the safety and effectiveness of surgical innovations. Experimenting with new techniques on animals and then trying them out on patients has served as the dominant model through which surgeons have developed new surgical procedures. Clinical experience, intuition, and pathophysiologic rationale—the very methods that EBM proponents have sought to replace by standardized evidence based on scientific research—have long reigned as the key factors in clinical decision-making for surgeons.

Not surprisingly, the vast majority of surgical procedures have never been evaluated in randomized double-blind clinical trials. Many surgeons have contested the philosophical basis of following clinical trial principles to evaluate new surgical innovations. Unlike the "sugar pills" used in pharmaceutical clinical trials, a comparable "placebo control" for testing a surgical intervention often involves significant risk to the research subjects—especially when the "control" involves administering fake operations to patients. Mimicking the conditions of the actual treatment as closely as possible, sham surgeries are used in clinical trials to maintain double-blinding and minimize potential bias from patient and researcher expectations.

The few cases in which researchers have employed sham controls to test the efficacy of surgical interventions have been fraught with controversy. For example, American researchers during the 1990s studied the effectiveness of transplanting human embryonic dopamine neurons into the brains of patients with Parkinson's disease. Attempting to follow the stringent demands dictated by clinical trials, some of these studies randomly assigned patients to undergo either sham surgery or transplantation (Freed et al. 2001; Freeman et al. 1999). Participants in both groups received local anesthesia, had four holes drilled through the frontal bone of the skull, and had imaging studies performed on them (Freed et al. 2001). Although they underwent risky and invasive procedures that mimicked the actual transplantation procedure, patients in the control group did not get the actual injection of embryonic tissue.

The controversial deployment of sham surgery controls has incited much debate among North American doctors and bioethicists (Miller 2003). In an article entitled "I Need a Placebo Like I Need a Hole in the Head," Charles Weijer, a Canadian bioethicist and physician, critiques the "scientistic bias" of clinical trials: "The clinician's office is not a laboratory any more than a research subject is a lab rat. Clinical care and human response to disease are simply too complex to be captured by such a simplistic model" (2002, 71).

On the other side, supporters of surgical RCT have argued that these sham operations can be ethically acceptable as long as the benefits outweigh the risks of conducting the clinical trial—but who benefits and who is at risk in these determinations? In a 2005 survey of Parkinson disease clinical researchers, 97 percent of respondents believed that sham-surgery controls were better than unblinded controls from a scientific perspective for testing the efficacy of neurosurgical interventions. Furthermore, the survey designers noted that "half of the researchers believe an unblinded control efficacy trial would be unethical because it may lead to a falsely positive result" (Kim et al. 2005, 1357). In other words, a significant percentage of these researchers believed that sham surgery controls were not only scientifically superior but also ethically superior. But the problem with these abstract determinations of scientific and ethical superiority is that they fail to spell out for whom these interventions would be more scientific and more ethical—the participants in the control arm of the study getting a hole drilled in their heads? Future patients who would have evidence-based data on efficacy? Researchers attempting to advance their careers? The following section examines how these debates have played out in the context of experimental fetal cell transplantation procedures performed at the New Century clinic.

Mismatching Temporalities

Chinese neurosurgeon Huang Hongyun, the founding director of the New Century Neurosurgery Clinic, has faced critical scrutiny from experts at American and European medical institutions. Huang rose through the ranks of the military medical system during the 1980s and received his doctorate in neurosurgery from the Postgraduate Medical College of the Chinese People's Liberation Army (Zhongguo Renmin Jiefang Jun Junyi Jinxiu Xueyuan) in 1991.[3] After conducting postdoctoral research on spinal cord injury at New York University and Rutgers (the state university of New Jersey) in the late 1990s, Huang returned to Beijing in 2001 to advance his clinical career. Leveraging the basic science research he conducted on rat models in the United States, Huang immediately began utilizing olfactory ensheathing glial cells in experimental treatments for Chinese patients suffering from spinal cord injuries and neurodegenerative disorders.[4] News of his experimental procedure spread quickly in both domestic and foreign media, leading to a long waiting list of patients interested in the procedure (during the period I conducted my fieldwork, I documented patients from over eighty different countries who received Huang's fetal cell transplantation procedure). The fees he received from his clinical work funded his clinical and basic science experiments—a process that flipped the bench-to-bedside clinical trial model dominant in the United States.

Nonplussed by his unorthodox pathway, Huang's American critics have deployed the sociopolitical logics of the EBM movement to dismiss his experimental practice, contesting his credibility on the grounds that he failed to perform a randomized controlled trial with defined entry criteria and blinded, predefined outcomes using reliable and relevant measurement tools. Charging Huang with "misleading patients with ALS [amyotrophic lateral sclerosis]," neurologists at Columbia University's Eleanor and Lou Gehrig MDA/ALS Research Center have argued:

> Clinics that expose patients to experimental treatments for ALS with inadequate scientific justification or the construct of a well-designed clinical trial do a disservice to the ALS community. They put patients at risk without sound evidence that the procedures could be beneficial, they do not contribute to scientific understanding of the therapy, and they divert resources away from legitimate research, thereby delaying the development of truly effective therapies. Lacking in acceptable safety standards and sufficient evidence for the potential to provide benefit, these clinics perform procedures that fail to meet the standards for ethical research set forth in the Nuremberg Code and Declaration of Helsinki. (Chew et al. 2007, 316)

Claiming both the scientific and moral high ground, these American researchers charged New Century clinicians with gross scientific and ethical negligence. Their invocation of the Nuremberg code implied that the Chinese fetal cell transplantation procedure could be compared to the horrific experiments perpetrated by Nazi physicians on concentration camp prisoners in the name of advancing medical research.

But Huang and his advocates have inverted the objections of these detractors by arguing that the experimental fetal cell procedure provided a more ethical alternative to the clinical trial model of his American critics. With the support of patients and their families, Huang has challenged conventional clinicians for focusing too much on preliminary animal studies while neglecting patients who are suffering or dying now. This temporal challenge has resonated in particular with people diagnosed with ALS, whose life demands do not match up with research timelines. A former golf pro from Florida who underwent fetal cell transplantation in November 2004 described an ALS research symposium he had attended back home, which featured leading neurologists describing promising therapies in the pipelines. An audience member asked how soon these treatments would be available. The speaker replied with a lengthy explanation about setting up clinical trials, noting that it would probably take about five years to make the necessary preparations for ensuring a quality study. The ALS patient interrupted him to declare that he was talking to a room of dead people.

Not surprisingly, as a review of "medical progress" for ALS published in the *New England Journal of Medicine* observed, "The lack of effective treatment has caused many patients and their families to become activists, raising money for research and bypassing traditional granting agencies" (Rowland and Shneider 2001, 1696). Describing these patient-led efforts as "guerrilla science," the authors nevertheless urged that "such approaches must first be attempted in animals to evaluate their safety and efficacy" (Rowland and Shneider 2001, 1696–1697).

But faced with a catastrophic illness, many ALS patients are not willing to wait for the conventional course of biomedicine to wind its way through years of animal testing, double-blind clinical trials, and paper publishing before a vetted treatment becomes available. The ALS patients coming to the New Century Neurosurgery Clinic want help now, because they worry that they may no longer be around even a year from now. Next month they may no longer be able to use their legs. The month after that they may no longer be able to use their hands. Sooner or later, they will lose their ability to communicate. And by the end, they will no longer be able to breathe on their own.

Echoing critics of the research timelines for anti-HIV/AIDS drugs (Epstein 1996), ALS patients argue that since they are already facing a "death sentence," they should have the choice to engage in "guerrilla science" and pursue experimental therapy. Many of them are thus willing to subject their bodies to relatively risky experimental procedures in the hopes of staving off their unrelenting disease. In an email newsletter Stephen Byer circulated in June 2004 to other families contemplating the experimental Chinese surgery, this father of an American patient captured the prevailing attitude among those who made the decision to undergo the fetal cell transplantation:

> Dr. Huang has not performed mouse or other lab animal tests, feeling, as many of us do, that such tests are not of direct benefit to humans, particularly those with a notably rapid advancing disease such as ALS. He regularly speaks of the questionable practice of extensive lab animal testing while people are dying and appears far more concerned with saving lives than proving theories on mice.

Huang's patients have championed him for placing them at the center of his work rather than concentrating on animal models of disease. But American critics have disparaged this focus on treatment over rigorous testing, noting that "such evasions are a classic mark of the charlatan" (Judson 2005). While American scientists see Huang's focus on alleviating suffering as evidence of quackery, patients see this as a badge of his dedication to their cause—and this divergence in emphasis demarcates the essential difference between clinical trials and experimental treatment.

Huang and many of his patients have explicitly rejected the principle of randomized placebo-controlled trials for surgical interventions. Whether taking place in China or the United States, clinical research has been plagued by the problem of therapeutic misconception (Sankar 2004; Henderson et al. 2007). Despite filling out laborious consent forms that have been scrutinized word-by-word by institutional review boards, patients participating in a clinical trial often believe—erroneously—that the trial will benefit them directly. That clinical investigators are often doctors creates even more confusion. Although the ultimate aim of the trial may be to benefit the patient population, the particular patients participating in the trial have been selected to further the aims of science rather than to improve their actual condition. Thus, some patients may receive nothing at all, while others may receive dosage levels known to be ineffectual. The focus is on testing the drug, not treating the patient. This logic becomes even more fraught when shifting from pharmaceutical trials to testing a surgical intervention on a person facing a devastating illness.

From an ALS patient's point of view, he or she cannot afford to get a placebo or a known ineffectual dose—even less so in the case of receiving a sham surgery for control purposes. Involvement in such a study literally asks the patient to sacrifice him or herself for the greater good with no personal benefit whatsoever, and even take on significant risks in the case of testing a surgical intervention. This might be more comforting if there had historically been more progress on ALS treatment. Despite the thousands of research papers that have been published about ALS, however, neurologists have produced very little in terms of concrete results for patients. As an American ALS patient receiving treatment at New Century Hospital noted, in a clinical trial "you don't even know if you get the stuff or not. You may be helping other people but you don't know if you're helping yourself." To subject vulnerable patients to the risks of surgery and administer only saline injections seemed particularly problematic to Huang and his supporters. While swallowing a sugar pill might pose little physical risk to patients, getting holes drilled into the skull was an entirely different matter.

Huang was incensed when I asked him about the criticism he received for not submitting his procedure to the rigors of a clinical trial. He shouted that a "sham surgery control" (*kongbai duizhao shoushu*, literally, "blank comparison surgery") was "completely against Chinese medical ethics":

> Let's say if some country permitted the use of sham surgical controls for the sake of so-called scientific ethics, I would completely despise that country! How could medical ethics have fallen to such an extent? To treat a person no longer as a human, to treat a patient not as a human, but instead to treat them as a machine or animal that we can conduct experiments on? This is very immoral! I think this type of vile medical practice should be eliminated! . . . To disregard a patient's wounds, to

disregard the possible dangers and side effects, which perhaps may even be life-threatening—to disregard all of this! And all to achieve what? So-called scientific conclusions!

Huang was outraged by the suggestion that he should cut open a group of patients and inject only saline in their brains or spines in the name of science. Present-day discussions of clinicians injecting their patients with saline solution (in the name of scientific inquiry) oddly echo Lao She's satire of the 1930s entrepreneurs injecting hapless patients with hypodermic needles filled with jasmine and longjing tea. But in Huang's rendering, the American clinicians pushing for sham surgical controls are the medical quacks. Huang inverted the logic of his American critics, declaring sham surgical procedures to be morally reprehensible. He adamantly refused to conduct clinical trials in which a portion of his patients would be subjected to the dangers of a craniotomy without receiving the fetal cell transplantation.

Generating a Paper Trail

Although Huang rejected the practice of sham surgery as unethical, this did not mean that he abandoned medical science. Having witnessed his efforts over the past several years, I would argue that Huang has tried to carry out "evidence-based" research with the available resources at his disposal—albeit with abbreviated timetables and protocols that aspired to ideal scientific principles but often fell victim to expedient concerns. Since the very start of his fetal cell transplantation odyssey in 2001, Huang has sought to document, substantiate, and track the results of his pioneering experimental surgery. In the following pages, I examine the methodologies that Huang and his colleagues have deployed to substantiate their findings.

Since 2001, Huang has published more than thirty scientific papers in Chinese- and English-language journals on the specific type of fetal cell deployed in his experimental treatment: olfactory ensheathing glial (OEG) cells. The form, content, and methodology of these papers have changed over time in response to his expanding audiences. Huang's first OEG paper—a three-page laboratory report written in Chinese—appeared in the academic journal of his home hospital on his return to China (Huang et al. 2001). Although this initial paper was based on laboratory experiments on animal models during a postdoctoral fellowship at Rutgers, Huang's return to clinical employment at the Naval General Hospital meant increased pressure to translate the previous years of basic science research he had been conducting in the United States into more immediate and practical applications for use in a Chinese clinical setting. Given the prevailing medical consensus on the inability of the adult mammalian brain and spinal cord to regenerate after damage, what explained these lab rats' unexpected functional improvements? Based on a post-mortem examination of the spinal cord tissue of the rats that had demonstrated functional recovery, Huang reported evidence of axon regrowth across the injured area (Huang et al. 2001, 66). This was the holy grail of spinal cord injury research, which had slowly begun to challenge the longstanding dogma on the failure of central nervous system (CNS) regeneration (Horner and Gage 2000). Huang's brief paper did not include details of his immunostaining process or visual evidence of the rats' regenerated axons to substantiate his remarkable claim. But for Huang, who was at heart a clinician interested in the practical significance of his research efforts, the rats' improved functional ability was the crucial proof of the experiment's potential. With an eye toward his next steps, Huang concluded that "OEG cells are the better candidate for promoting nerve regeneration compared with other types of cells with regenerative capacities, holding very good prospects for application in clinical trials" (Huang 2001, 67).

Huang followed this basic science paper with his first clinical case report the following year, describing a diverse group of twenty-three patients with spinal cord injuries he had treated with OEG cell transplantation at Navy General Hospital (Huang et al. 2002). Averaging these patients' pre- and postsurgical status scores, Huang claimed that this first group of treated patients showed statistically significant improvements in both motor and sensory function. He published a much larger case series report a year later, documenting the results of OEG transplantation for 171 Chinese spinal cord injury (SCI) patients (Huang et al. 2003a). By this point, Huang had orchestrated his move to the civilian Chaoyang Hospital, the main teaching hospital affiliated with Beijing's Capital Medical University. He inaugurated his move by publishing this report in the academic journal of his new institution.

Huang's Chinese-language clinical reports attracted the attention of domestic journalists, who seized on the potential for transforming the lives of patients with feature reports entitled "There Is Hope for Spinal Cord Injury Treatment" (CCTV 2002) and headline news declaring "Paraplegic Patients Back on Their Feet Is No Longer a Dream" (Sha 2003). These television and newspaper reports, among others (see Liu 2002; Tang et al. 2002; Zhao 2002), generated a waiting list of patients numbering into the thousands from throughout China. But because Huang's initial clinical reports were published in Chinese by the local journals of the hospitals that employed him, their localized nature rendered them largely inaccessible to those outside China, including the international scientific community.

In October of 2003, Huang published his first English-language scientific paper on OEG transplantation, in which he reported the results of the experimental transplantation on 171 patients with spinal cord injury (Huang et al. 2003b). The data were identical to the Chinese-language report he had published in the Capital Medical University's academic journal earlier that year (Huang et al. 2003a). But because this version was written in English and published in a journal indexed in the Science Citation Index as well as MEDLINE, Huang suddenly began receiving an upsurge of interest from outside China. The publication of the paper also coincided with the first U.S. patients to undergo the OEG transplantation procedure.

Neuroscientists, clinicians, and journalists from the United States and Europe began boarding planes to Beijing, skeptical but eager to witness the purportedly groundbreaking experimental procedure. These visiting delegations included neurologists from the Miami Project to Cure Paralysis, representatives from the U.S. Embassy, the scientific journal *Nature*'s Asia-Pacific correspondent, and the *New England Journal of Medicine*'s editor-in-chief, among many others. Despite attracting deep skepticism, Huang's alleged achievement of overturning medical dogma (that the central nervous system was incapable of regenerating) was also what drew the attention of these outside experts. As Mei Zhan (2001) has demonstrated in a different context, traditional Chinese medical practitioners also leveraged "miracle-making" abilities to negotiate the complex relationship between marginality and fame, promoting not just their own individual careers but also developing broader translocal communities. Huang utilized his experimental biomedical procedure to assert clinical superiority, unabashedly proud that he as a Chinese neurosurgeon was treating foreign patients who had been written off by their own clinicians as incurable.

Threatened by Huang's claims of clinical superiority, various clinicians from the United States, Spain, and the Netherlands began to perform and publish independent assessments of patients who had undergone the Beijing fetal cell transplantation procedure, dismissing Huang's results as biased and potentially flawed in execution (Chew et al. 2007; Dobkin, Curt, and Guest 2006; Giordana et al. 2010; Guest, Herrera, and Qian 2006; Piepers and Van den Berg 2010). After examining the pre- and postsurgical status of a seventy-year-old American woman with ALS who had undergone the Beijing OEG transplantation procedure, the Columbia University neurologists discounted Huang's efforts as unscientific on the following grounds:

> There is no evidence of defined enrollment criteria, a control group, blinded outcome measurements, or systematic collection of safety data, and no unbiased method has been used to discern positive or negative

> effects. These procedures cannot be considered credible research because they lack standardized impartial assessments of safety and efficacy. (Chew et al. 2007, 316)

Citing only two papers written by Huang, the Columbia University neurologists dismissed Huang's reports as anecdotal publicity rather than scientific research. This assessment overlooked the twenty-four additional research papers Huang had published between 2001 and 2006 on the safety and feasibility of OEG transplantations, including a clinical case series report on the results of eighty-eight ALS patients who had received OEG transplantation (Huang et al. 2006b), a study on the interim safety of OEG transplantation for ALS based on magnetic resonance imaging (MRI) (Chen et al. 2006a), and two studies on the long-term safety of OEG transplantation based on three-year follow-up results assessed by MRI (Chen et al. 2006b; Huang et al. 2006a).

Since many of these overlooked papers were published in Chinese-language scientific journals, the Columbia clinicians' failure to account for them is understandable, if not excusable. Furthermore, the data from Huang's earlier papers could arguably be considered "unsystematic" since he did not establish a control group for comparison and selected participants based on a convenience sample of available patients. But it is wrong to accuse Huang of failing to establish the safety and feasibility of his procedure. A demonstration of safety does not require a randomized controlled trial. Indeed, a simple registry tracking patient outcomes would be sufficient for documenting adverse events. The Columbia clinicians' criticism of Huang makes a fundamental error: it conflates the documentation of safety with methods for establishing efficacy. Calling Huang unscientific is also a rhetorical cheap shot against a Chinese physician who has made clear attempts to document and study his results. These flawed critiques ultimately reveal the unequal power dynamics characterizing global biomedical research.

From Anecdote to Data

Although he realized that he was operating within an unequal playing field, Huang was determined to address critics who dismissed his results as mere "anecdotal reports" and "unsubstantiated clinical practice" (Dobkin, Curt, and Guest 2006, 5, 13). The New Century neurosurgery staff adamantly rejected sham surgery protocols, but they strove to demonstrate that systematic data on safety and efficacy could be collected in other ways (Chen et al. 2010; Huang et al. 2008). They devised several alternative methodologies to produce quantifiable data on the short- and long-term outcomes for their experimental fetal cell procedure.

In order to provide an objective basis for comparing functional change, for example, New Century clinicians adopted internationally recognized assessment tools such as the American Spinal Injury Association Impairment Scale (ASIA). Initially developed in 1982 by American researchers attempting to achieve greater consistency and reliable data among centers participating in the National SCI Statistical Center Database, these standards were endorsed by the International Medical Society of Paraplegia in 1992 and have been updated multiple times by a committee of international experts (Kirshblum et al. 2011). The ASIA classification standards thus provided an internationally recognized, consistent method for categorizing the level of motor function and sensory impairment following spinal cord injury. Based on the conventional understanding that damaged neurons in the central nervous system cannot be repaired or renewed, a person's ASIA classification was not supposed to change once his or her condition stabilized. The New Century staff's careful documentation of their patients' ASIA classifications before and after the OEC transplantation procedure thus served as a key strategy for procuring potential proof of the experimental surgery's effectiveness.

For ALS patients, New Century clinicians utilized the ALS Functional Rating Scale, which calculated patients' quality of function in four main areas: gross motor tasks, fine motor tasks, bulbar functions (e.g., speech and swallowing), and respiratory function (ALS CNTF Treatment Study Phase I–II Group 1996; Cedarbaum et al. 1999). Specific questions addressed a patient's ability to perform tasks related to speech, salivation, swallowing, handwriting, cutting food, dressing and hygiene, turning in bed, walking, climbing stairs, and breathing. While the original scale was designed "to track progression of patients' disability in ALS" (Cedarbaum et al. 1999, 18), Huang's staff adapted the scale to assess the possibility of improvement following the fetal cell transplantation.

In addition to using existing methods for assessing patients' neurological condition, New Century clinicians also developed their own assessment tools to document key changes they observed in patients receiving the fetal cell therapy. One of the crowning achievements of Dr. Alan, the chief Chinese neurologist, was to develop the New Century Hospital Spinal Cord Injury Daily Life Functional Rating Scale (NCHSCI-DLFRS). The NCHSCI-DLFRS measures functional outcomes that affect patients' everyday lives, including the ease of accomplishing activities such as eating, grooming, dressing, transferring to a bed, and using the toilet. The New Century scale also measures physiological sensations neglected by the ASIA scale, such as muscular tension, ability to sweat, and also degree of pain—issues that contribute to, or even define, a patient's quality of life.

A patient's physiological status and quality of life could thus be quantified through these existing and new scales, enabling New Century clinicians to measure whether a given patient had experienced overall improvement following the fetal cell transplantation. The quantification of functional change also facilitated New Century clinicians' efforts to make comparisons between patients and construct extensive databases that could then be translated more easily into scientific publications and the production of proof that these surgical interventions worked. With strings of numbers rounded to the nearest hundredth of a decimal, complete with confidence intervals and p-values, the New Century physicians attempted to shield themselves against foreign critics' allegations of unscientific methodology. These numbers were generated from worksheets labeled with acronyms such as ASIA, ISNCSCI (International Standards for Neurological Classification of Spinal Cord Injury), NCHSCI-DLFRS, GMFM (Gross Motor Function Measure), ICARS (International Cooperative Ataxia Rating Scale), and CP-ADL (Cerebral Palsy Activities of Daily Living) as Huang and his staff expanded their treatment protocol to encompass additional neurological disorders. From a semiotic perspective, the proliferation of technical jargon, precise measurements, and acronyms in New Century's research papers and medical reports operated as a key strategy for bolstering the Chinese clinicians' scientific credibility (Latour and Woolgar 1986; Hyland 2010).

Tracking Chinese Patients

Although the New Century clinicians devoted months to perfecting these various scales and measurement tools, their biggest challenge was finding patients willing to undergo these assessments. Tracking long-term postoperative outcomes for Chinese patients was a Sisyphean endeavor, not just for Huang but for many of the Chinese clinicians I spoke with in hospitals across Beijing. As an internist at Peking Union Medical College Hospital explained to me, patients from throughout the country came to Beijing in search of treatment at the nation's top hospitals, often as a last resort after treatment at their local and regional level hospitals had failed. Once they left the capital city (carrying their medical records back with them to their hometowns), these patients generally had no additional contact with their Beijing-based clinicians. There were few incentives on either side to remain in touch. The busy schedules of clinicians such as the Peking Union internist left little time in the workday to track down patients who had already received treatment. Patients who had already spent significant sums seeking treatment did not want to waste additional money on failed therapies or extraneous consultations. Furthermore, unequal resource distribution in rural and urban health institutions made it difficult to compare results and measure efficacy. Many of the clinicians I spoke with in Beijing distrusted not just the diagnostic conclusions but even the laboratory tests reported by colleagues based at provincial hospitals. Carrying out clinical trials or even basic follow-up care was thus a difficult proposition in Beijing's clinical context.

Unlike the foreign patients, few of Huang's Chinese patients communicated directly with the head neurosurgeon. Most of them had signed up for the fetal cell transplantation through the typical bureaucratic channels at previous hospitals where Huang had operated and went through the central hospital registration system before being transferred to Huang's neurosurgery department for treatment. Huang's institutional moves impeded his ability to access the medical charts of patients treated at these previous hospitals. He thus had to resort to other means of obtaining data, such as persuading colleagues at his former institutions to conduct chart reviews for him with the enticement of co-authoring joint research publications. Huang also tried to reconstruct his own set of patient charts. Starting in 2005, he began inviting former Chinese patients to return to Beijing for a free follow-up assessment. So even though Huang had decided to focus almost exclusively on treating foreign patients at New Century Hospital, the daily neurosurgical rounds generally included a few Chinese patients.

Brother Lu, a businessman from a county-level city in Zhejiang province who had suffered a spinal cord injury several years ago, came to New Century Hospital for a follow-up assessment in March of 2006. A slight, lively man in his early thirties with short cropped hair, Lu was a veteran of the experimental fetal cell therapy, having undergone the procedure at Naval General Hospital in 2002 in the hopes of bringing movement and sensation back to his paralyzed lower body. At Huang's invitation, he had returned to Beijing with his new wife, a vivacious woman who eagerly attended to his needs. Lu described the arduous process of getting from his home in Zhejiang to the capital city. The train ride alone took thirty hours, a difficult journey even for a man with a top-of-the-line wheelchair accompanied by a doting wife. Prior to the train ride, Lu and his wife had taken an overnight bus from his hometown to get to the nearest major train station. Embarrassed that the clinic was already paying for everything else, they presented only the receipts for the train to the hospital staff for reimbursement.

Lu explained that he had been seriously ill after being injured in a car accident in May of 2000. At the time, he was unable to breathe on his own and had to rely on a ventilator. He slowly regained his ability to breathe during a three-month stay at the local hospital, but he remained unable to move his hands. He later saw a Chinese Central Television broadcast about Huang's surgery for spinal cord injury patients. Although the news report did not mention any specific details, he eagerly called the station and asked them for the name of the hospital where the surgeries were taking place. Describing the fetal cell transplantation he subsequently underwent in Beijing, he told me that the conditions were incredibly crowded and that he had to make many of the arrangements himself during his previous trip to the capital. Lu seemed quite pleased that he did not have to "dig out money from his own pocket" for the follow-up assessment. He was impressed by Huang's new facilities at New Century Hospital and even inquired about the possibility of undergoing a second fetal cell transplantation.[5]

Lu laughed about being too thin and showed me how his arm muscles had atrophied. He was missing the end of his middle finger and noted that his hands were weak. But, he explained, before the fetal cell surgery he was not able to move them at all. Now he could feed himself, but he still had trouble gripping things and could not write. As he scratched his head with his left hand, the head nurse Teacher Yan remarked that his hand seemed quite good and that he was doing very well compared with many of the other seriously injured patients they treated. Lu agreed: "In the past I could only lie on the bed watching the ceiling all day long. I could not sit up; whenever I sat up I would feel dizzy and lightheaded, see golden stars in my eyes, feel nauseous, and vomit. After the surgery, I can now sit in a wheelchair for more than two hours and push myself around the entire yard."

Cheerful and upbeat, Brother Lu was clearly thriving in spite of his paralysis from the waist down. He talked animatedly about owning a successful factory that employed twelve workers and produced tripods. He also gestured proudly at his pretty wife, declaring that they had gotten married two years after his successful fetal cell transplantation procedure. Lu's wife and his postsurgical sexual functioning later became the subject of much gossip among the New Century staff, as they discussed how a disabled man was able to convince such an attractive woman to marry him.

Following Up: An Epic Cross-Country Journey

While Brother Lu was one of a few dozen Chinese patients who returned to Beijing for a free follow-up assessment at New Century Hospital, hundreds more did not make the onerous trip to Beijing. Despite Huang's offer to pay for all of their expenses, many other former patients failed to respond even after three rounds of invitations. Dr. Alan, the chief neurologist, speculated that some of the patients probably did not believe that the department would actually shoulder the costs, since this deviated from standard operating procedure among Chinese medical institutions. Those unhappy with the results of the surgery could hardly be expected to waste their time by coming back.

In order to fill in the gaps on their patient follow-up spreadsheet, Huang decided that New Century doctors would go out to visit the patients in their own homes. During the spring of 2006, the staff began preparations for a three-month "follow-up" (*suifang*) expedition across eastern China, with the goal of "using the shortest route to interview the most patients." The goal of this suifang journey was to perform follow-up assessments on at least 100 Chinese patients. This seemed like a modest figure, given the nearly 200 Chinese patients who had received OEG fetal cell transplantations at Naval General between 2001 and 2002 and over 300 patients at Chaoyang Hospital between 2002 and 2004. But the task of locating former patients scattered across the country turned out to be a nearly insurmountable challenge.

The neurosurgical trainees spent their off-duty hours poring over a wall-sized map of the country to plot out potential routes. In reviewing patient records, they discovered that most of the listed addresses were not detailed enough. Some entries listed only the town name, while others neglected to include the apartment number. The office staff thus devoted their energies to calling former patients to obtain more precise street addresses and alert them that two doctors would be coming to give them a free check-up. In the process, they also discovered that a significant portion of the listed phone numbers were no longer in service. This was not surprising, as visitors from other provinces often avoided racking up long-distance charges by buying local cell phone cards to use during their sojourn in Beijing, which they would then discard on returning home.

While the office staff struggled to plot out a viable route, the nurses put together a patient education pamphlet that featured translated testimonials from foreign patients describing their improvements after surgery. The small handbook also included advice about physical therapy and exercise routines to help patients strengthen their atrophied muscles. Head Nurse Sophie explained that following the fetal cell procedure, patients needed to exercise in order for there to be any effect. Particularly for spinal cord injury patients, "the less you move, the worse it gets." Sophie noted how foreigners invented all sorts of contraptions to help themselves and praised their resourcefulness. She remarked that many of her foreign patients still had such positive attitudes despite their serious conditions, in contrast to some of the rural Chinese patients who tended to lie in bed doing nothing after getting sick or injured. The pamphlet was thus designed to "inspire Chinese patients to work harder."

Huang's chief neurologist, Dr. Alan, would be in charge of doing the follow-up neurological assessments (*pingfen*, literally "assigning points") while Dr. Gu would drive the vehicle and operate the video camera to create a visual record of postoperative changes. The deputized doctors spent several weeks poring over car advertisements and visiting dealerships—already a favorite pursuit among the clinic doctors, who were swept up in the nationwide car craze seizing China's increasingly wealthy urban residents. Envisioning difficult driving conditions outside the capital, Dr. Alan initially wanted to buy a jeep to help navigate the dirt roads along which many rural patients lived. Dr. Gu preferred something larger and more enclosed, anticipating that they might have to sleep in the vehicle overnight in more isolated areas without roadside inns. After heated debates over the merits of domestic versus foreign manufacturers, they ended up purchasing a Hyundai minivan manufactured by the Chinese state-owned Jianghuai Automobile Group.

Eager to witness their data collection practices in action, I tried unsuccessfully to persuade the doctors to take me with them on their suifang journey. Invoking urban stereotypes about the "backward" (*luohou*) and "uncivilized" (*bu wenming*) character of rural residents, Dr. Alan explained that it was "not safe" (*bu anquan*) for a young woman to go on the trip. Dr. Gu agreed, noting that outside the city, the situation could be "very chaotic" (*hen luan*). Although they framed the issue in terms of my personal safety, I caught the underlying subtext that my presence as a woman with a foreign passport would be too much of an inconvenience for two men traveling in a van all summer. I thus had to make do with tracking the doctors' progress retrospectively, through intermittent text messages and email updates during the times they had access to internet cafes.

The challenges Drs. Alan and Gu faced in trying to track down their Chinese patients illustrate the broader difficulties in obtaining credible evidence for determining the effectiveness of experimental treatments in contemporary China. Despite the detailed mapping endeavor prior to their journey, Dr. Alan reported significant problems in locating patients. Even in cases where the staff had tracked down detailed addresses, the doctors in their Hyundai minivan discovered houses that were marked for demolition, street names that had been changed, and even entire neighborhoods that had been razed to make way for new development projects. Suspicious of these outside doctors' intentions, local work units and public security bureaus were often not willing to cooperate. As with the mismatch in temporalities I discussed earlier between patient lives and research timelines, these challenges in locating patients also mark a stark difference between the temporality of social and economic change in contemporary China in relation to the temporality of illness.

Of the patients they did manage to locate, Dr. Alan told me that many of them were disappointed by the outcome of the surgery: their high hopes for the experimental treatment did not match up with the actual results they had experienced. A few were lucky to live near county-level rehabilitation hospitals and continued to work on building finger dexterity, standing up, achieving independence, or broader life goals. But many more were bedridden, relying on busy family members and friends to help them accomplish basic tasks and routines. Dr. Alan was not surprised by these disheartening results. Without access to the personal trainers and physical therapy equipment that many of Huang's foreign patients took for granted, Chinese patients lacking resources and support were perhaps doomed to experience deteriorating muscle mass and worsening neurological function.

Conclusion

In this chapter, I have examined how Huang and his staff have responded to allegations of quackery while devising new standards of evidence. Although Huang contested the ethics of sham surgery, this did not mean that he abandoned scientific principles. Since pioneering the fetal cell transplantation procedure in 2001, Huang and his colleagues have sought to document, substantiate, track, and assess the results of this experimental surgery. Having tracked these efforts for over a decade, I would conclude that the neurosurgeons and neurologists at New Century Hospital have made credible efforts to develop alternative forms of "evidence-based" clinical practice with the available resources at their disposal— but these attempts have had mixed results.[6]

In urging Chinese citizens to "seek truth from facts" (*shi shi qiu shi*), both Mao and Deng quoted this proverb to highlight pragmatism as a guiding principle. As the different challenges posed by conducting follow-up studies with both domestic and foreign patients illustrate, New Century's research timetables and protocols aspired to utopian scientific principles but often fell victim to dystopian realities. Establishing efficacy under the controlled conditions of an optimized clinical trial fails to account for the practical constraints shaping the everyday lives of clinicians, patients, and their families. In the case of Chinese patients, the problems Drs. Alan and Gu faced in trying to track them down illustrate the broader difficulties in obtaining reliable evidence for determining the effectiveness of experimental treatments in contemporary China. These difficulties in locating patients expose the fault lines created by socioeconomic change as they intersect with the spatial and temporal dynamics of illness and injury. Paralyzed patients have been "left behind" in China's interior hinterlands, with little access to assistive technologies and rehabilitation services. The precariousness of their situation marginalizes them even further as ambitious Chinese clinicians such as Huang turn toward more promising candidates able to comply with research protocols. As I have documented elsewhere (Song 2017), the New Century staff ironically found that they could produce better data and results with their foreign patients, who simultaneously offered a more lucrative financial return. Despite living tens of thousands of miles away across oceans and entire continents, foreign patients tended to respond quickly to emails and were often eager to provide updates of their postoperative status. Internet-mediated communication enabled Huang and his staff to conduct more robust follow-up studies on their foreign patients and document the longer-term outcomes of their experimental procedure. By exposing the competing interests and differential outcomes embedded in New Century's fetal cell therapies, I ultimately challenge not just how we measure the efficacy of experimental medicine but also the unintended consequences generated by the fraught nexus of state-market-technoscience in a transnational world.

NOTES

Adapted from *Biomedical Odysseys: Fetal Cell Experiments from Cyberspace to China* by Priscilla Song. Copyright © 2017 by Princeton University Press. Reprinted by permission.

1. Chinese regulations for clinical trials have become even more stringent than the U.S. Food and Drug Administration's guidelines—at least on paper. See, for example, People's Republic of China National People's Congress (2001) for a drug administration law adopted in 1984 and revised in 2001.

2. These concerns harken back to a long tradition of anthropological scholarship on magic, science, and religion, which has engaged questions of rationality and efficacy in evaluating the status of knowledge claims produced by other societies (Evans-Pritchard 1976; Good 1994; Malinowski 1948; Nader 1996; Tambiah 1990).

3. See chapter 5 of *Biomedical Odysseys* (Song 2017) for a more detailed account of Huang's biography.

4. An olfactory ensheathing glial (OEG) cell is a type of nerve support cell that has shown potential in laboratory and clinical trials for restoring functional recovery after spinal cord injury.

5. Chinese patients were expected to procure the aborted fetus(es) for the experimental procedure. Despite his relative wealth and satisfaction with his earlier surgery, Lu found this requirement to be too onerous to undergo a second OEG fetal cell transplantation.

6. This chapter examined the scientific and ethical debates engendered by Huang's experimental procedure during the initial years of his neurosurgical enterprise. Since 2009, Huang has focused on transforming neurological repair from conditional experiment to standard medical practice by establishing the field of "neurorestoratology." See Song (2017) for an analysis of this subsequent phase of knowledge production.

References

ALS CNTF Treatment Study (ACTS) Phase I–II Study Group. 1996. "The Amyotrophic Lateral Sclerosis Functional Rating Scale: Assessment of Activities of Daily Living in Patients with Amyotrophic Lateral Sclerosis." *Archives of Neurology* 53: 141–147.

Bynum, W. F. 2006. "The Rise of Science in Medicine, 1850–1913." In *The Western Medical Tradition 1800 to 2000*, edited by W. F. Bynum, Anne Hardy, Stephen Jacyna, Christopher Lawrence, and E. M. Tansey, 111–229. New York: Cambridge University Press.

CCTV—*See* China Central Television.

Cedarbaum, Jesse, Nancy Stambler, Errol Malta, Cynthia Fuller, Dana Hilt, Barbara Thurmond, and Arline Nakanishi. 1999. "The ALSFRS-R: A Revised ALS Functional Rating Scale That Incorporates Assessments of Respiratory Function."

[BDNF ALS Study Group (Phase III).] *Journal of the Neurological Sciences* 169: 13–21.

Chen, Lin, Hongyun Huang, Yancheng Liu, Haotai Xi, Feng Zhang, Jian Zhang, Hongmei Wang, Chengqing Gou, Ruiwen Liu, Chao Jiang, Zhao Jiang, Zixing Xie, and Chunyan Luo. 2006a. "Xiù qiào xìbāo yízhí zhìliáo jī wěisuō cè suǒ yìnghuà zhèng zhōngqí ānquán xìng píngjià" [Interim Safety Evaluation of Fetal Olfactory Ensheathing Cell Transplantation for Amyotrophic Lateral Sclerosis]. *Zhōngguó Línchuáng Kāngfù* [Chinese Journal of Clinical Rehabilitation] 10(25): 24–26.

Chen, Lin, Hongyun Huang, Hongmei Wang, Haitao Xi, Chengqing Gou, Jian Zhang, Feng Zhang, et al. 2006b. "Xiù qiào xìbāo yízhí zhìliáo wǎnqí jǐsuǐ sǔnshāng de cháng qī ānquán xìng píngjià: Cí gòngzhèn chéngxiàng sān nián suífǎng" [Long-Term Safety of Fetal Olfactory Ensheathing Glial Cell Transplantation in Treatment of Malignant Spinal Cord Injury: A Three-Year Follow-Up with Magnetic Resonance Imaging]. *Zhōngguó Línchuáng Kāngfù* [Chinese Journal of Clinical Rehabilitation] 10(5): 28–29, 32.

Chen, Lin, Hongyun Huang, Haitao Xi, Zihang Xie, Ruiwen Liu, Zhao Jiang, Feng Zhang, et al. 2010. "Intracranial Transplant of Olfactory Ensheathing Cells in Children and Adolescents with Cerebral Palsy: A Randomized Controlled Clinical Trial." *Cell Transplantation* 19(2): 185–191.

Chew, Sheena, Alexander G. Khandji, Jacqueline Montes, Hiroshi Mitsumoto, and Paul H. Gordon. 2007. "Olfactory Ensheathing Glia Injections in Beijing: Misleading Patients with ALS." *Amyotrophic Lateral Sclerosis* 8: 314–316.

China Central Television (CCTV). 2002. "Jǐsuǐ sǔnshāng zhìliáo yǒuwàng" [Hope for Spinal Cord Injury Treatment]. *Jiànkāng Zhī Lù* [Road to Health Program]. February 4.

Cochrane, Archibald L. 1972. *Effectiveness and Efficiency: Random Reflections on Health Services*. London: Nuffield Provincial Hospitals Trust.

Craig, Sienna. 2012. *Healing Elements: Efficacy and the Social Ecologies of Tibetan Medicine*. Berkeley: University of California Press.

Dobkin, Bruce H., Armin Curt, and James Guest. 2006. "Cellular Transplants in China: Observational Study from the Largest Human Experiment in Chronic Spinal Cord Injury." *Neurorehabilitation and Neural Repair* 20(1): 5–13.

EBM Working Group—*See* Evidence-Based Medicine Working Group.

Epstein, Steven. 1996. *Impure Science: AIDS, Activism, and the Politics of Knowledge*. Berkeley: University of California Press.

Evans-Pritchard, E. E. 1976. *Witchcraft, Oracles, and Magic among the Azande*. Oxford: Clarendon Press.

Evidence-Based Medicine Working Group. 1992. "Evidence-Based Medicine: A New Approach to Teaching the Practice of Medicine." *Journal of the American Medical Association* 268(17): 2420–2425.

Freed, C. R., P. E. Greene, R. E. Breeze, W. Y. Tsai, W. DuMouchel, R. Kao, S. Dillon, et al. 2001. "Transplantation of Embryonic Dopamine Neurons for Severe Parkinson's Disease." *New England Journal of Medicine* 344(10): 710–719.

Freeman, T. B., D. E. Vawter, P. E. Leaverton, J. H. Godbold, R. A. Hauser, C. G. Goetz, and W. Olanow. 1999. "Use of Placebo Surgery in Controlled Trials of a Cellular-Based Therapy for Parkinson's Disease." *New England Journal of Medicine* 341: 988–992.

Giordana, Maria, Silvia Grifoni, Barbara Votta, Michela Magistrello, Marco Vercellino, Alessia Pellerino, Roberto Navone, et al. 2010. "Neuropathology of Olfactory Ensheathing Cell Transplantation into the Brain of Two Amyotrophic Lateral Sclerosis (ALS) Patients." *Brain Pathology* 20(4): 730–737.

Good, Byron. 1994. *Medicine, Rationality, and Experience.* New York: Cambridge University Press.

Guest, James, L. P. Herrera, and T. Qian. 2006. "Rapid Recovery of Segmental Neurological Function in a Tetraplegic Patient Following Transplantation of Fetal Olfactory Bulb-Derived Cells." *Spinal Cord* 44: 135–42.

Guyatt, G., J. Cairns, D. Churchill, et al. 1992. "Evidence-Based Medicine: A New Approach to Teaching the Practice of Medicine." *Journal of the American Medical Association* 268(17): 2420–2425.

Harris, Gardiner, and Walt Bogdanich. 2008. "Drug Tied to China Had Contaminant, F.D.A. Says." *New York Times*, March 6. http://www.nytimes.com/2008/03/06 /health/06heparin.html.

Henderson, Gail, Larry Churchill, Arlene Davis, Michele Easter, Christine Grady, Steven Joffe, Nancy Kass, et al. 2007. "Clinical Trials and Medical Care: Defining the Therapeutic Misconception." *PLoS Medicine* 4(11): e324. doi:10.1371/journal. pmed.0040324.

Higgins, Julian, and Sally Green, eds. 2011. *Cochrane Handbook for Systematic Reviews of Interventions.* Version 5.1.0. The Cochrane Collaboration. http://handbook .cochrane.org/.

Horner, Philip, and Fred Gage. 2000. "Regenerating the Damaged Central Nervous System." *Nature* 407(6807): 963–970.

Huang, Hongyun, Lin Chen, Hongmei Wang, Bo Xiu, Bingchen Li, Rui Wang, Jian Zhang, et al. 2003a. "Niánlíng duì xiùqiào xìbāo yízhí zhìliáo jǐsuǐ sǔnshāng liáoxiào de yǐngxiǎng" [Age Influences on Recovery Outcome of Spinal Cord Injury Treated by Intraspinal Transplantation of OECs]. *Shoudu Yike Daxue Xuebao* [Journal of Capital University of Medical Sciences] 24(1): 56–59.

Huang, Hongyun, Lin Chen, Hongmei Wang, Bo Xiu, Bingchen Li, Rui Wang, Jian Zhang, et al. 2003b. "Influence of Patients' Age on Functional Recovery after Transplantation of Olfactory Ensheathing Cells into Injured Spinal Cord Injury." *Chinese Medical Journal* 116(10): 1488–1491.

Huang, Hongyun, Lin Chen, Hongmei Wang, Haitao Xi, Chengqing Gou, Jian Zhang, Feng Zhang, et al. 2006a. "Xiùqiào xìbāo yízhí zhìliáo wǎnqí jǐsuǐ sǔnshāng ānquán xìng píngjià 38 gè yuè cí gòngzhèn suífǎng jiéguǒ" [Safety of Fetal Olfactory Ensheathing Cell Transplantation in Patients with Chronic Spinal Cord Injury: A 38-Month Follow-Up with MRI]. *Zhongguo Xiufu Chongjian Waike Zazhi* [Chinese Journal of Reparative and Reconstructive Surgery] 4: 439–443.

Huang, Hongyun, Lin Chen, Hongmei Wang, Jian Zhang, Feng Zhang, Yancheng Liu, Haitao Xi, et al. 2006b. "Xiùqiào xìbāo yízhí zhìliáo jī wěisuō cè suǒ yìnghuà zhèng: 88 Lì jìnqí jiéguǒ bàogào [Olfactory Ensheathing Cell Transplantation in the Treatment of Amyotrophic Lateral Sclerosis: Recent Result Report of 88 Cases]. *Zhongguo Linchuang Kangfu* [Chinese Journal of Clinical Rehabilitation] 10(1): 39–41.

Huang, Hongyun, Lin Chen, Haitao Xi, Hongmei Wang, Jian Zhang, Feng Zhang, and Yancheng Liu. 2008. "Fetal Olfactory Ensheathing Cells Transplantation in Amyotrophic Lateral Sclerosis Patients: A Controlled Pilot Study." *Clinical Transplantation* 22: 710–718.

Huang, Hongyun, Kai Liu, Wencheng Huang, Zonghui Liu, and Wise Young. 2001. "Xiùqiào xìbāo cùshǐ jǐsuǐ cuò liè shāng hòu shénjīng zàishēng hé gōngnéng huīfù de yánjiū" [Olfactory Ensheathing Glias Transplant Improves Axonal Regeneration and Functional Recovery in Spinal Cord Contusion Injury]. *Haijun Zong Yiyuan Xuebao* [Naval General Hospital Journal] 14: 65–67.

Huang, Hongyun, Hongmei Wang, Bo Xiu, Rui Wang, Ming Liu, Lin Chen, Shubin Qi, et al. 2002. "Xiùqiào xìbāo yízhí zhìliáo jǐsuǐ sǔnshāng línchuáng shìyàn de chūbù bàogào" [Preliminary Report of Clinical Trial for Olfactory Ensheathing Cell Transplantation Treating Spinal Cord Injury]. *Haijun Zong Yiyuan Xuebao* [Naval General Hospital Journal] 15: 18–21.

Hyland, Ken. 2010. "Constructing Proximity: Relating to Readers in Popular and Professional Science." *Journal of English for Academic Purposes* 9(2): 116–127.

Institute of Medicine. 2009. *Initial National Priorities for Comparative Effectiveness Research.* Washington, DC: National Academies Press.

Judson, Horace Freeland. 2005. "The Problematical Dr. Huang Hongyun." *MIT Technology Review.* http://www.technologyreview.com/news/405327/the -problematical-dr-huang-hongyun/.

Kim, Syh, S. Frank, R. Holloway, et al. 2005. "Science and Ethics of Sham Surgery: A Survey of Parkinson Disease Researchers." *Archives of Neurology* 62: 1357–1360.

Kirshblum, Steven, William Waring, Fin Biering-Sorensen, Stephen Burns, Mark Johansen, Mary Schmidt-Read, William Donovan, et al. 2011. "Reference for the 2011 Revision of the International Standards for Neurological Classification of Spinal Cord Injury." *Journal of Spinal Cord Medicine* 34(6): 547–554.

Lao, She. 2011. *Kāi Shì Dàjí: Lǎo Shě Xiǎoshuō Jīngdiǎn* [A Brilliant Beginning: Classic Short Stories by Lao She]. Beijing: Beijing Yanshan Chubanshe.

Latour, Bruno, and Steve Woolgar. 1986. *Laboratory Life: The Construction of Scientific Facts.* Princeton, NJ: Princeton University Press.

Liu, Li. 2002. "Huáng Hóngyún: Gěi Sāng Lán men dài lái xīwàng" [Huang Hongyun: Giving Hope to Sang Lan]. *Kējì Rìbào* [Science and Technology Daily], August 26, 8.

Löwy, Ilana. 2000. "Trustworthy Knowledge and Desperate Patients: Clinical Tests for New Drugs from Cancer to AIDS." In *Living and Working with the New Medical Technologies: Intersections of Inquiry*, edited by Margaret Lock, Allan Young, and Albert Cambrosio, 49–81. New York: Cambridge University Press.

Malinowski, Bronislaw. 1948 [1922]. *Magic, Science, and Religion and Other Essays.* Westport, CT: Greenwood Press.

Marks, Harry. 1997. *The Progress of Experiment: Science and Therapeutic Reform in the United States, 1900–1990.* New York: Cambridge University Press.

Miller, F. G. 2003. "Sham Surgery: An Ethical Analysis." *American Journal of Bioethics* 3: 41–48.

Nader, Laura, ed. 1996. *Naked Science: Anthropological Inquiry into Boundaries, Power, and Knowledge.* New York: Routledge.

Piepers, Sanne, and Leonard van den Berg. 2010. "No Benefits from Experimental Treatment with Olfactory Ensheathing Cells in Patients with ALS." *Amyotrophic Lateral Sclerosis* 11: 328–330.

Rentz, E. Danielle, Lauren Lewis, Oscar J. Mujica, et al. 2008. "Outbreak of Acute Renal Failure in Panama in 2006: A Case-Control Study." *Bulletin of the World Health Organization* 86(10): 737–816.

Rothman, David. 1997. *Beginnings Count: The Technological Imperative in American Health Care.* New York: Oxford University Press.

Rowland, Lewis P., and Neil A. Shneider. 2001. "Medical Progress: Amyotrophic Lateral Sclerosis." *New England Journal of Medicine* 344(22): 1688–1700.

Sankar, Pamela. 2004. "Communication and Miscommunication in Informed Consent to Research." *Medical Anthropology Quarterly* 18(4): 429–446.

Sha, Wenru. 2003. "Jiétān bìngrén chóngxīn zhànlì bù zài shì mèng" [Paraplegic Patients Back on Their Feet Is No Longer a Dream]. *Zhōngguó Xiāofèizhě Bào* [China Consumer News], April 10.

Song, Priscilla. 2017. *Biomedical Odysseys: Fetal Cell Experiments from Cyberspace to China.* Princeton, NJ: Princeton University Press.

People's Republic of China National People's Congress. 2001. "Zhōnghuá rénmín gònghéguó yàopǐn guǎnlǐ fǎ" [Drug Administration Law of the People's Republic of China].

Tambiah, Stanley J. 1990. *Magic, Science, Religion, and the Scope of Rationality.* Cambridge, UK: Cambridge University Press.

Tang, Xianwu, Yuxiao Wang, and Er Ji. 2002. "Tā shìzhe ràng lúnyǐ rén zhàn qǐlái" [He Attempts to Let People in Wheelchairs Stand Up]. *Kējì Rìbào* [Science and Technology Daily], January 29, 1.

Thompson, Charis. 2013. *Good Science: The Ethical Choreography of Stem Cell Research.* Cambridge, MA: MIT Press.

Tröhler, Ulrich. 2005. "Quantifying Experience and Beating Biases: A New Culture in Eighteenth-Century British Clinical Medicine." In *Body Counts: Medical Quantification in Historical and Sociological Perspectives*, edited by Gerald Jorland, Annick Opinel, and George Weisz, 19–50. Montreal: McGill-Queen's University Press.

Weijer, Charles. 2002. "I Need a Placebo Like I Need a Hole in the Head." *Journal of Law, Medicine and Ethics* 30: 69–72.

Yao, J. L. 1996. "Perinatal Transmission of Hepatitis B Virus Infection and Vaccination in China." *Gut* 38(Supplement 2): S37–S38.

Zhan, Mei. 2001. "Does It Take a Miracle? Negotiating Knowledges, Identities, and Communities of Traditional Chinese Medicine." *Cultural Anthropology* 16(4): 453–480.

Zhao, Shaohua. 2002. "Tānhuàn bìngrén shì zěnyàng zhàn qǐlái de" [How Do People with Paralysis Stand Up?]. *Jiànkāng Shíbào* [Health Times], January 17. http://www.people.com.cn/GB/paper503/5238/549351.html.

DIVERGENT TRUST AND DISSONANT TRUTHS IN PUBLIC HEALTH SCIENCE

Katherine A. Mason

Xu Dan had had it. She shuffled some papers around and then threw them down with disgust and swung her chair around to face me. Shaking her head, as if she had just finished scolding a small child, she told me, "You know I am trying to get at the TRUTH [in English], but it's impossible to do that in this context. . . . These statistics are totally meaningless—and the district CDCs don't care about the truth, only I seem to care." It was the summer of 2009, and Xu had a deadline coming up for submitting her year-by-year comparative statistics for the incidence of hand foot and mouth disease in Tianmai, a large city in the Pearl River Delta region where she worked at a local, city-level Center for Disease Control and Prevention (CDC) and where I was in the midst of conducting thirteen months of ethnographic fieldwork (see Mason 2016a). Xu had collected all of the necessary numbers and she was on track to easily make her deadline. The leaders would be pleased: the incidence of hand foot and mouth disease in Tianmai, according to the numbers submitted to her by her colleagues at the district- and "street"- (*jiedao*) level CDCs, had fallen year by year. This had, by all accounts, been an unusually light summer for reports of the disease. Xu would likely be praised for her report, and the work was simple and straightforward.

But Xu was anything but pleased. As a part-time PhD student at a university in Hong Kong, and with a master's degree in epidemiology from a British university, Xu was worldly, ambitious, and well versed in what she sometimes referred to (in English) as "real science." And what she was currently engaged in, she explained to me, was not real science. The numbers on hand foot and mouth that she had received from the lower level CDCs in the public health hierarchy were obviously too low. This in part could be attributed to the attention and manpower that had been diverted to addressing that summer's outbreak of pandemic H1N1 influenza—"no one is paying attention to hand foot and mouth right now," she told me. From Xu's perspective the deeper problem, however, was that her colleagues didn't really "care" about the veracity of the numbers. Xu told me,

> It's so frustrating because I know that what I'm producing is not the "real truth" [in English]. I do statistics, but they're not real statistics. I want scientific truth, but the lower levels don't care about science— they don't get it when I say if we compare hand foot and mouth disease between this year and last year, we need to know that they are always reporting, consistently, in both years. I know they [the statistics reported] are not the truth, so it's depressing, because I have to produce a report regardless, have to do this stuff, it's my job and it's what the leaders expect. But I know it's not the truth, not science, nothing to do with reality, so "*meiyou yisi, dou meiyou yisi*" [it's meaningless, it's all meaningless].

On the surface, said Xu, she shouldn't care either. "What I do, what it has to do with the truth or anything is not related to that," she told me. "My salary is the same whether I do something or not, whether there's an outbreak or not, whether I stop the outbreak or not." And yet even though she insisted that it mattered neither to her colleagues nor to her job whether she got at the "truth" or not, it mattered deeply to her.

Xu contrasted the "truth" that she sought with the "correct" data that fulfilled the obligations of a *guanxi* relationship—the reciprocal personal relationships that constitute one of the foundations of Chinese sociality (Fei 1992). As I have described elsewhere, "correct" data do not necessarily correspond to what biomedical scientists might consider "reality" (see Mason 2016b). For example, reporting to a *guanxi* partner a 100 percent vaccination rate for a vaccination campaign that actually vaccinated only 80 percent of the target population would not be a "true" statement, but it would be a "correct" one.

Xu found that most of her colleagues—particularly those who were older and not as well educated as she was—were steeped in the expectations of producing information that was "correct." They were therefore uncooperative in producing information that Xu and her similarly young and educated peers considered to be "true"—that is, corresponding to an objective, immutable reality that existed independently of human engagement. Science studies scholars have long argued that such a reality is unattainable. As Lisa Gitelman and Virginia Jackson (2013) put it, all "raw" data is in fact already "cooked" simply by virtue of being collected by humans (see also Fleck 1979; Latour and Woolgar 1986). And yet, Xu saw the insertion of human concerns into the production and dissemination of data as an ethically problematic act.

Lorraine Daston and Peter Galison, in their historical study of the concept of "objectivity," suggest that to be "objective" in this moralized sense, "is to aspire to knowledge that bears no trace of the knower" (2007, 17). In Xu's view, personal relationships and the obligations therein rendered data inherently nonobjective and therefore neither true nor moral. Even when there was no distraction like the H1N1 outbreak to get in the way, she told me, most Chinese scientists did not know how to, nor did they care to learn how to, be objective. Xu attributed this characteristic both to a lack of competence (traceable to a lack of sufficient biomedical education) and to a deep moral failing. Even if they were properly trained in statistical methods, Xu felt that she could not *trust* her colleagues to produce the "real" truth. Too deeply embedded in personal relationships to remain objective, too concerned about pleasing leaders or saving face to seek any truth that might cast themselves or their superiors in a negative light, and too dependent on social networks to approach science with the kind of purity and "innocence" (*danchun*) that characterized the work of foreign (Western) scientists, Chinese scientists were, in Xu's estimation, doomed to forever produce an inferior or even false form of scientific knowledge.

Saving China—and the World—through Science?

As I describe elsewhere (Mason 2016a), the roots of Xu's discontent—if not the problem itself—can be traced back to a seminal event in the history of public health, and public health science, in China. In early 2003, a novel flu-like virus called SARS (severe acute respiratory syndrome) swept through the Pearl River Delta region where Xu worked, before crossing the border into Hong Kong and then traveling around the world. More than 8,000 people contracted SARS globally before its disappearance in July of that same year, and the disease had a fatality rate of about 10 percent. More than half of the cases and nearly half of the deaths were recorded in China. After at first denying the scope of the outbreak, the Chinese central government admitted in April 2003 that the disease was out of control and instituted a series of aggressive measures to contain it, including mass quarantines and closures of public spaces.

The global health community, led by the World Health Organization, both blamed China for the initial outbreak and praised it for the rapid containment that followed. While it exposed grave insufficiencies in a public health system that had fallen into disrepair during thirty years of economic reforms, the SARS incident also presented Chinese scientists with an opportunity. Suddenly, the country's and the world's attention was on public health in China, and large amounts of money from both domestic and foreign sources began to flow to local-level public health departments to help rebuild infrastructure and hire personnel. In particular, reform efforts were geared toward the biomedicalization and professionalization of local CDCs—especially those located in the Pearl River Delta region, where SARS had originated.

Just prior to the SARS outbreak, the central government had begun the process of transforming thousands of local Mao-era Anti-Epidemic Stations (AES), which focused primarily on local sanitation and vaccination work, into local CDCs, which were to be centers for public health management and biomedical research at the provincial, city, district, and county levels. However, this project initially lacked funding or a clear mandate for what the CDCs were supposed to do that was different from what AESs had done before. SARS lent urgency and a sense of mission to the reform process. A longtime member of the Tianmai CDC explained it to me this way: "After SARS, that made clear what the goal of the Tianmai CDC is. Now primarily what we do is . . . gradually building preparations, contingency for sudden public health incidents, as well as some infectious disease response." A big part of the move toward infectious disease control was the establishment of a sophisticated online system of disease reporting, which was intended to make Chinese public health data more transparent and decrease the degree to which it was "cooked" prior to being shared (Gitelman and Jackson 2013; Mason 2016b).

Along with a shift in focus from basic sanitation to emergency preparedness, the CDCs after SARS also invested heavily in scientific research. Post-SARS public health science was portrayed as "saving" China in two senses: better science would prevent another SARS-like outbreak that could threaten both individual lives and the international standing of the nation-state; and better science would promote the reputation and standing of China's public health community throughout the world, establishing its members as trustworthy professionals and global citizens. It was primarily in pursuit of this second goal that Xu and many of her contemporaries arrived at the Tianmai CDC throughout the early and mid-2000s.

The educational requirements for employment at city- and district-level CDCs in the big urban centers of the Pearl River Delta rapidly became more stringent after the SARS outbreak. At the Tianmai CDC, a vocational degree (*dazhuan* or *zhongzhuan*) was all that was required to be hired prior to SARS; afterward a bachelor's and eventually a master's degree was required, with a PhD expected for some positions. This sudden change in expectations resulted in a rapid shift in the demographic profiles of those engaged in local public health work. Whereas local AESs had been dominated by career bureaucrats in their late forties and fifties, most of those who entered the local CDCs in the 2000s were in their twenties and thirties and considered themselves to be scientists (*kexuejia*) and professionals (*zhuanye renyuan*).

Among the approximately sixty young city- and district-level CDC members between the ages of twenty-two and forty whom I interviewed in the urban Pearl River Delta during my fieldwork, I found a consistent pattern of initial optimism followed by bitter disappointment in the scientific enterprise in which they were engaged. For example, shortly after she arrived at the Tianmai CDC, Ke Jin, a young woman in her mid-twenties with a master's degree in microbiology, told me that she greatly admired how the leaders of the CDC "saw opportunity" in the outbreaks they investigated. She said, "They would go deal with the problem, and at the same time they would think, 'Oh, how can I get a paper out of this?' or 'What can I invent and sell that would help us do this better? What *keti* [grant] can I apply to that's related? What research can I do on this topic?' They have really limber minds [*naozi feichang linghuo*]." But for Ke, as for many of her colleagues, the optimism did not last. When I visited Tianmai for follow-up research two years later, Ke had left to take a job at a provincial-level CDC, from where, one of her friends told me, she hoped to go abroad for further study. "She found it too tiresome here," the friend told me. "She couldn't do the science she wanted to do."

Trust and Truths

One of the main reasons for the bitterness that my interlocutors eventually developed toward their work was, they said, that too many of their (primarily older) colleagues were "not scientific enough" (*bugou kexue*) and therefore, as Xu Dan had said, not worthy of trust. Rather than "following the data" as Xu and others believed a good scientist should, these other colleagues continued to follow the requirements both of center leaders (*lingdao*) and of their complex networks of social relationships. They were thus frequently faced with pressure to produce numbers that would reflect favorably on those leaders and networks. And yet from the perspectives of most of the older CDC members I knew, the system of interpersonal relationships that were required to accomplish this actually provided the primary foundation on which trust was built and out of which their moral codes were drawn. For this older generation, the relationship obligations that Xu found so antithetical to establishing trust was precisely what was required for trust.

The rituals that shaped *guanxi* interactions in the public health setting—primarily in the form of banqueting among officials representing different local CDCs and other public health institutions—continued to be necessary for any local public health project to proceed even after SARS.[1] At the heart of these rituals was the building and maintenance of *renqing*, or "human feelings" (Yan 1996; Yang 1994). According to Arthur Kleinman, *renqing* is "both social and deeply personal; it captures the dialectical quality of experience; it is individual and interpersonal. It represents the moral core of experience" (1995, 111). The existence of *renqing* in a *guanxi* relationship allowed trust to be built; without it, little could be accomplished.

Even the young newcomers I knew relied on *renqing*, built through *guanxi* rituals, to carry out any sort of public health effort. But while *renqing* remained critically important in practice, the definition of what sort of public health data constituted the "truth" had shifted after SARS. The types of truths that *guanxi* partners had always been trusted to produce were no longer consistent with the types of truths that the new public health professionals expected. The process and the product were the same, but the classification and judgment of this process and product had shifted. The young newcomers to public health after SARS redefined "correct" truths as poor or even falsified data; in many cases "correct" truths no longer had any truth value for my interlocutors at all. As a result, those who produced them no longer had the moral authority that they once did.

In his classic work, the *Social History of Truth*, Steven Shapin shows how for scientists in seventeenth-century England, the making of scientific truths was dependent on a form of deep professional trust. Scientists felt comfortable believing and building on the work of their colleagues because they trusted that they were sharing in certain practices, procedures, and moralities as a basis for producing science—and because they trusted the specific people doing the work. They trusted precisely because the data they received from each other were *not* raw data floating in the ether, but rather were highly contextualized data steeped in a human relationship between producer and receiver. While a certain amount of cynicism has since come to undermine this trust in contemporary scientific communities, Shapin argues, collegial trust remains a cornerstone of (Western) scientific knowledge production.

In Shapin's examples, people trusted scientific data when they trusted the particular people who were producing it and handing it over. They trusted those people to actively participate in producing good data and *not* to keep themselves and their relationships removed from this process. They trusted them because they knew them, and they knew that they were going to behave in what they thought of as a "professional" manner—that they were going to do their best to produce good quality data.

The case of Chinese public health science after SARS provides a fruitful node from which to reassess and reimagine the role of trust and truth in the making of scientific data. On the one hand, *personal* trust between colleagues remained the basis for all scientific knowledge production in China—one could not hope to gather data or publish a paper without drawing on trusted networks of *guanxi* "friends." On the other hand, what many Chinese scientists perceived as a lack of what I have elsewhere called *professionalized trust* made it difficult for this mode of trusting to endure the changes that were reshaping public health institutions in the years after SARS (Mason 2016a). For these scientists, a reliance on personal trust made professionalized trust hard to achieve.

The worldly and ambitious young people who flooded local CDCs after SARS sought to replace, rather than make use of, interpersonal trust based in *guanxi*. Their concept of the professionalized trust they hoped to build in its stead was quite different from Shapin's concept. While Shapin found that trust within the scientific profession was embedded in relationships between specific people doing the work, and in the context of those particular personalities and their particular expertise or track records, Xu and her colleagues found in this familiarity only a basis for *dis*trust. Instead, they insisted that professionalized trust had to be anonymous and transferable to anyone by means of their status, disconnected from the particularities of their backgrounds. Professionalized trust was to be based on one's status as *a* colleague but equally importantly on one's lack of status as any *particular* colleague. "You should just do it because you are a scientist, not because you are friends," Fu Qiang, a Tianmai CDC member in his mid-twenties explained. The anonymity and nonspecificity were important, precisely because the opposite was so important for producing the "correct" kind of truth that the new generation of Chinese public health scientists wanted to discard.

Convinced of an incompatibility between interpersonal relationships and the production of "real" truth, young public health scientists turned to foreign colleagues for examples of a science that they imagined to be somehow free from personal relationship considerations and therefore purer and more representative of what they thought of as "reality." In this way they tended to see foreign scientific communities, particularly those in the United States, as more "fair." As Professor Luo, a former Tianmai CDC employee and researcher at a university in Guangzhou who had previously spent several years in the United States as a postdoctoral fellow, told me, "China's system is much less fair and objective. For example, in the United States, if you apply for an NIH [National Institutes of Health] grant, you'll get a specific number assigned—say, 84.5. If you have 84.5 and someone else has an 84, then you definitely are ranked ahead of them. It's very scientific, very objective. But here it's totally subjective. . . . It's all *guanxi*." Like Xu and Fu, Luo conjured an incorruptible Western objectivity that Shapin and most other science studies scholars would insist does not exist.

This fantasy of objectivity and fairness formed the backbone of what my interlocutors dubbed "international science" (*guoji kexue*). "International science" for them was located in what Gieryn (2002) refers to as a "place-less" "truth-spot"—a vaguely "Western" sphere that does not need to be described but from which the truth is understood to emanate. This version of science sat at the pinnacle of a global science hierarchy that Chinese public health scientists were attempting to scale through collaborations, training abroad, and education—but which they felt doomed to be able to imitate only poorly if at all. Even on those occasions on which they felt they were able to produce a reasonably objective truth, my interlocutors viewed the science they produced as inferior because it was "not fresh" (*bu xinxian*). Several young people at the Tianmai CDC complained to me that in their scientific work the best they could do was either produce a "Chinese version" of studies already done abroad—for example by adding questions about green tea to a cancer epidemiology survey developed in the United States—or contribute to collaborative studies by providing access to large numbers of research subjects (Mason 2015). "What we're good at in China is that we can get a lot of data very fast," Xu explained, when describing to me a collaborative study in which she was participating that involved surveying several thousand migrant workers over the course of several weeks. "International" scientists from the United States, Europe and Australia whom I met during my fieldwork who were participating in collaborations like this one largely accepted this imbalance of power, asserting in conversations with me that they were aware that "Chinese data is really bad" as one professor put it and that, nevertheless "the Chinese are really good at rounding up research subjects." As they saw it, their role was to educate and to set an example of, as one Australian put it, "how to do good science."

"Good-Enough" Truths

One of the more puzzling aspects of this global science hierarchy was not so much why the inequality in relationships existed or why "international" scientists accepted Chinese evaluations of their science as purer and superior. Much of this can be explained by the hubris of the Western scientists I met, who, while acknowledging that their data were also imperfect, had no problem assuming that these imperfect data were more accurate than anything their Chinese colleagues were likely to produce. The more interesting question for me became why, exactly, the younger generation of Chinese scientists came to trust Western scientists whom they had not ever met to deliver a "truth" that they never would expect from their closest Chinese peers. How did loyalty and reciprocity come to be associated with dishonesty and falsehoods, and cold distance with honesty and truth?

Perhaps one answer lies in the nature of who and what, exactly, were being trusted. The trust that my interlocutors afforded to foreign scientists was not so much trust in their own willingness to produce truth so much as it was trust in their ability to remove themselves from the equation and be "objective" (2007). That is, Xu, Luo and other young Chinese scientists *trusted foreign scientists to know that they could not and should not trust themselves.* This is quite different from the trust afforded to peers in the scientific networks that Theresa MacPhail (2014) describes, in which the active participation of the partner is key in establishing the truth value of the information. According to MacPhail, if the information is entirely detached from its source, its truth value cannot be assured; instead, a trusted partner needs to intervene in the process of conveying that truth by assessing and asserting the veracity of the truth being delivered.

Because they saw their work environments as so antithetical to the production of objective truth, Chinese public health scientists sought to at least position themselves as trusted partners to foreign scientists. But this was difficult to achieve, in part because many of my interlocutors found themselves at a crossroads between two definitions of truth that were incommensurate with each other but across which they had to translate.

For example, as an English-speaking, biomedically educated professional who had spent time abroad, Xu served as a bridge between the world of Chinese public health science and the world of what she called "international science." Xu, Ke, and their similarly situated young colleagues were often called upon to arrange and manage international collaborations, to translate articles written by colleagues into English and format them properly to be submitted to English language journals, to host foreign visitors, and in general to translate the research and "everyday work" (*richang gongzuo*) that they did into the kind of science that was legible to (Western) outsiders. While doing my fieldwork, I noticed that each department of the Tianmai CDC had at least one highly educated young person who served as what I came to think of as a kind of foreign collaborator ambassador, charged with hosting any foreign visitors and dealing in any international correspondence.

At the same time, in their home institutions these cultural and linguistic ambassadors were expected to follow the same rules as everyone else with regard to respecting the requirements of *guanxi* relationships and avoiding offending crucial *guanxi* partners, without whom it continued to be difficult or impossible to do anything. They could not effect a full-scale revolution in how science was done at home, and yet they also felt they could not directly present what their home institution's science produced to their international collaborators in ways that the collaborators would find acceptable—convinced as they were of a Western commitment to an objective truth. Their collaborators reinforced this perception with their frequently vocal skepticism of the veracity of Chinese statistics.

As a result, Chinese public health scientists were stuck producing what might be termed "good-enough" truths. Discussing a research project in which she discovered that some of her Chinese collaborators had partially fabricated large swaths of data rather than admit that they could not produce what she had requested (an exercise in the production of "correct" data that would once have been considered all the "truth" that was necessary), Xu told me, "I have to complete the project, so I will have to just use the data I have." Doing her best to work with data that she strongly felt did not represent the "truth," Xu nonetheless attempted to maintain her own status as a trustworthy scientist by manipulating that data in as statistically sound, scientifically accurate, and ultimately "truthful" a manner as she felt was possible. In these data manipulations she transformed "correct" data into a good-enough truth that was not ideal but was at least the product, she felt, of some sort of proper scientific method. The numbers themselves may have been what her international collaborators would think of as unscientific or even false, but by subjecting those numbers to proper scientific manipulation—in this case, by engaging in careful biostatistical calculations that adhered closely to internationally accepted scientific methods—Xu felt that some sort of truth value could nevertheless be created (Mason 2016a).

And yet in producing her good-enough truths, Xu found herself perpetuating a cycle of mistrust, as she then became part of an apparatus that produced and disseminated what both she and the international colleagues she was trying to please considered to be poor and inaccurate data. In the process she both damaged relationships with those *guanxi* partners with whom she expressed increasingly open frustrations and ran up against persistent barriers to producing the kinds of relationships she fantasized about building with colleagues abroad.

Several of Xu's young colleagues also described a similar double bind. In redefining the provision of "correct" data by their *guanxi* partners as an immoral act, they suggested a radical departure from norms of loyalty and reciprocity that had long dominated the state-run scientific apparatus of which they were a part. Running any public health campaign or research project associated with a local CDC required the active building and maintenance of *guanxi* ties through entertainment activities such as banqueting (*yingchou*). At the heart of *guanxi*-based scientific cooperation were norms of loyalty and reciprocity, which stipulated that *guanxi* partners returned favors and provided their "friends" with what they asked for. If they asked for you to vaccinate 10,000 people and you were able to vaccinate only 5,000, then you would nonetheless report that you had vaccinated 10,000 because that was the "truth" that was desired, even if it did not match up with vaccinations being administered on the ground.

The new public health scientists, however, were trying to redefine what the obligations of *guanxi* actually were. This meant redefining the meaning of the "gift" that *guanxi* partners were supposed to give to each other (Yan 1996). Instead of providing a *guanxi* partner with evidence of perfect success in carrying out a project, a partner was now supposed to give the "gift" of what "international science" would consider the objective truth. But the older and less educated members of the local CDCs confessed to me that they did not really know how to do that—and even if they did, how could they trust that it would not come back to haunt them? If they turned over numbers that made their own institutions look bad, how could they be certain that their own boss would not become angry, or that the security of their job would not be affected?

Older public health workers at the local level frequently spoke with me of the high levels of anxiety that they experienced in trying to figure out what numbers to produce and report and to whom. "No matter what, someone will be angry," a fifty-two-year-old sanitation inspector who had worked in public health in Tianmai for decades told me. The rules of the game had been changed on him by one side but not by the other. The leaders at his own institution still expected to be provided with numbers that made them look good, while their partners at other institutions—some of whom were higher in the government hierarchy than their own leaders—now expected something different. What had been the epitome of a moral response suddenly became a potential act of deception. The most trusted of *guanxi* partners suddenly became the most distrusted of scientific colleagues. This created a good deal of tension between those for whom different meanings of truth applied.

"Discounted" Truths

Xu contrasted the good-enough truth that she strove to create with two other forms of truth, both related to the level of trust she felt she could or could not afford to the provider of the data. One was what I call "discounted truth." This was a form of truth that emerged from the principle articulated most famously by Mao Zedong: "The serious question at present is not only that the lower levels fabricate, but that we believe them. . . . Reported successes should be discounted, split 30/70—can three out of 10 be taken as fake, and the seven as true?" (quoted in Blum 2007, 122). Discounted truth was what a savvy reader of Chinese statistics would surmise was the likely "real" number hidden in the "correct" statistic provided by a trusted *guanxi* partner. The *guanxi* partner's job was to produce a correct statistic. His or her colleague's job was to interpret that statistic correctly

When Xu went about discounting the truths that she received, she applied a stepwise logic to the process, which she based on her estimation of the motivations and scientific backgrounds of those providing the data. The discount taken on the number provided depended on how much the person could be trusted to put at least some effort into collecting good data. If the person *duguo shu* (had some education), but perhaps had not spent time abroad and was unwilling to stray too far from the obligations of what was "correct," then a number provided of "10,000" cases of measles might be interpreted to represent a "real" number of about 15,000. This was the discounted truth. My interlocutors would assume that the number provided had been inflated (or deflated) in the direction of what was perceived to be politically palatable and desired by a *guanxi* partner, but that it was not radically different from the "real" number. The assumption was that the person had probably put some effort into collecting and processing the data and had the skills to do it. He or she respected "international science" and would want to provide something that wasn't too far off from the "real" truth. There would be some relationship between the data provided and the "real" truth.

On the other hand, if the person had no education, he or she could not be trusted to even put in that effort. If this type of colleague provided a number of 10,000, Xu explained that probably she would estimate the "real" number as something higher—say, 30,000. And if the person Xu was working with was, like herself, young, well educated, and fed up with the system, maybe the discounted truth would be closer to 10,000 still—although this also depended on the extent to which he or she was also answering to leaders that expected him or her to stick closer to "correct" than to "true." In other words, the discounted truth represented Xu's best guess as to what the "real" number hidden in the statistics she received was, based on how much she trusted the provider of the statistics to produce something that bore at least some relation to what she thought of as "reality." In telling me about a similar process that she used in interpreting numbers from those working under her at a district-level CDC, Dr. Guo explained, "to understand the numbers, you just really have to understand who the people [*renyuan*] are."

And yet what so frustrated Xu and Guo were the gaps between the good-enough truths that they attempted to produce based on the numbers provided, the discounted truths with which they processed those numbers, and the "imagined truths" that they felt certain that scientists elsewhere in the world were managing to produce. The imagined truth was that truth produced by a romanticized vision of the "international" scientist, whom they imagined to be objective, neutral, and devoid of the influence of interpersonal relationships. Trust was afforded to strangers who were deemed capable of removing themselves and their relationships from the business of producing truths.

The interesting thing about this stranger trust was that it was based in a fundamental skepticism and form of *mistrust*. It was the ability of international scientists to *mistrust themselves*—that is, to acknowledge that the insertion of any of their own ideals or concerns into the production of statistics would sully those statistics and detract from their ability to reflect a "real" truth—that made them seem trustworthy to others.

Several of my interlocutors told me that this trust-producing form of mistrust extended to an ability and willingness to mistrust colleagues as well. Dr. Ying, a parasitologist at one of Tianmai's district CDCs, assumed that Western scientists would trust colleagues only if those colleagues were able and willing to remove themselves entirely from the data—thus effectively eliminating the need for interpersonal trust. They could be trusted because they knew not to trust, and because they did not expect anyone to trust them. Ying explained it this way: "Everybody is suspicious of everyone else. That's not good for friendship but it is good for science."

Thus a strong *distrust* of local colleagues, coupled with a nonreciprocal and highly romanticized trust of international collaborators—based paradoxically in those international collaborators' capacity for self-mistrust—produced an array of often dissonant "truths" for Chinese public health scientists. The result of all of this is that everyone was working with a number of truths at once, that these truths often contradicted each other, and that therefore expert knowledge gained from hands-on experience working within the Chinese public health system was needed to interpret and reconcile these various truths with each other. For example, in the case of disease reporting, if 100 cases were reported, depending on who did the reporting and who received the report, the discounted truth, good-enough truth, and imagined truth might all differ (with the "real truth" always existing somewhere beyond anyone's grasp).

The moral quandary for younger members of the local CDCs lay in their inability to produce the imagined truth, and the discomfort they felt working with the good-enough truth. In the end, young, highly trained scientists worried that they themselves did not actually deserve to be trusted any more than their colleagues, despite feeling that their intentions were better. They were stuck in a type of sociality that was fundamentally incompatible with their conception of good science.

Still, these young scientists wanted to believe that their education gave them a leg up in reaching their ideal of imagined truth—even if it hadn't been reached yet. For example, although Xu was implicated in the continued production and dissemination of faulty data, from her perspective *she* provided faulty data to her foreign collaborators because she had no choice, whereas many of her colleagues did so because of a moral lapse that allowed personal relationships to creep into an arena that should have been impersonal and objective.

In turning "correct" data into good-enough data for foreign partners, Xu also had to engage in a final stage of data manipulation. This was necessary, she insisted, because foreigners would not know how to properly interpret Chinese data.

In the interest of providing data that were as close to the "real" truth as she could get, Xu felt she had no choice but to insert her own judgments and opinions into what was being produced. She also knowingly participated in the transfer of what she considered to be poor data to those whose standards, she assumed, were higher. As a result of all this, she judged herself as ultimately untrustworthy. She did not trust herself to produce "real" truths. And she did not trust herself to step away from the data and present them as "raw," when she knew how "cooked" they were to begin with. In other words, she mistrusted herself, but for reasons that were different from those for which she felt a better scientist would (more productively) mistrust herself. "I know I am not a good scientist—at least not right now," she told me.

And yet Xu felt that if she could just get out and fight her way into becoming part of the international scientific community, then she would become trustworthy, too, because her raw materials would be such higher quality and would so much more closely resemble the "real" truth. She felt that she had the potential to be a trustworthy, neutral scientist—that she had the right character. She just didn't have the proper materials. And so this is also how she justified her own role in perpetuating the production and dissemination of truths that were not "real" truths. She had to do so, she said, in order to gain for herself and others like her the experience and connections that might in turn provide her with the opportunity to become a better scientist in the future.

The Innocent Foreigner

An important source of the trust that my interlocutors afforded to Western scientists was the trust that they had in the *institutions* with which these scientists were associated (see Carpenter 2010). In other words, as Niklas Luhmann (2000) put it, "system trust" preceded any particular form of "personal trust." Anthony Giddens (1990) argues that system trust is in turn dependent on the existence of face-to-face trust with representatives of the system whom he calls "access points." Most of those who subscribed the most strongly to trust in foreign institutions had experience with access points, thanks to stays abroad or interactions with visiting foreign colleagues. Importantly, these access points tended to consist of fairly brief and superficial encounters. Most commonly, public health scientists would spend a month or two abroad as visiting scholars, or would take part in welcoming foreign scientists to Tianmai for a conference or collaborative project.

Those who traveled abroad usually returned with tales of how foreign science was a world in which relationships didn't matter, and people went about producing knowledge without all the complicated (*fuza*) human relationships that made Chinese social life richer and also made Chinese science less truthful and scientific. In particular, "abroad" (*guowai*), as my interlocutors described it, was a place where life was "pure and innocent" (*danchun*).

One chilly afternoon in January 2009, for example, I was chatting with Dr. Li, a department head at the Tianmai CDC, about her ongoing diarrheal surveillance project. This was a project that she had adapted from a similar project with which she had been involved in Canada, where she had spent a brief period as a visiting scholar the previous year.

I asked Li what she thought about her time in Canada, and she said she found that life there was very "*danchun* [innocent]—you go to work and do your assigned things and then you go home—people are very *zhuanye* [professional]. . . . They are not always running around all the time like us." While she found this *danchun* lifestyle to be restful and relaxing—"a good atmosphere" that was conducive to "being scientific," she told me—many of her colleagues who had similar experiences found it to be stifling and dull. What everyone agreed on was that on the level of scientific knowledge production, the "innocence" of foreigners' lifestyles made it more possible for them to produce something close to the "real" truth.

The innocence that Dr. Li and others associated with *danchun* foreigners contrasted in interesting ways with the associations with professionalism that this innocence evoked. On the one hand, some of my informants almost infantilized foreign scientists as being innocent to the world and too childlike and unsophisticated to successfully participate in the complicated relationships that Chinese had to navigate. On the other hand, these same characteristics made foreign scientists seem more successful and respectable professionally. The ability to be innocent made it easier to seem trustworthy, because foreigners were not complex enough on the level of sociality to effectively deceive. My interlocutors' stories of living abroad gave the impression that, like good children unable to tell a lie, foreign scientists were not capable of the kind of conniving and manipulation that Chinese schooled in the art of *guanxi* were, and so my interlocutors did not need to know much about them to believe that what they were producing was, to the best of their abilities at least, the "truth." And because of the technical sophistication that mixed with that social innocence, their ability to produce the truth, should they so desire (which it was assumed they did), was, according to Li and others, quite high.

Thus foreign scientists were seen as technically and scientifically sophisticated but socially innocent, while Chinese scientists were portrayed as being socially sophisticated but scientifically immature. From this point of view, foreigners simply made better scientists, because good science, according to my informants' characterization, *should* be asocial. Therefore, even aside from their higher levels of education, access to better resources, and greater technical expertise, foreigners were seen as having a natural leg up in producing good science, due to their quiet, individualistic, and socially immature natures. By living a lifestyle so dull that one had nothing better to do but to work all day and sit quietly at home with one's family all night, these foreign colleagues demonstrated to my Chinese informants that personal relationships did not much matter to their work or in their lives (or at least, they thought this was what was being demonstrated), and therefore that the science they were producing was naturally closer to being purely "objective." This in turn led to a fatalism on the part of Chinese scientists like Xu, who despaired of Chinese scientists ever being able to disentangle themselves enough from their social networks to produce high quality science.

Daston and Galison describe what they call the "moralization of objectivity" this way:

> What unites the negative and positive sides of mechanical objectivity is heroic self-discipline on the one side, the honesty and self-restraint required to foreswear judgment, interpretation, and even the testimony of one's own senses; on the other, the taut concentration required for precise observation and measurement, endlessly repeated around the clock. It is a vision of scientific work that glorifies the plodding reliability of the bourgeois rather than the moody brilliance of the genius. It is also a profoundly moralized vision, of self-command triumphing over the temptations and frailties of flesh and spirit. (1992:83)

Although Daston and Galison were writing about the nineteenth-century European turn toward a fetishization of objectivity in science, much of this vision applies to how Chinese scientists in Tianmai similarly fetishized objective knowledge in early twenty-first century Euro-America. Foreign scientists were dull and even "plodding," yes, but that also made them, in this view, "pure," "innocent," and therefore, honest. In Li's view, because they were pure and innocent, so were their truths.

Although they admired this quality, both men and women who returned from visiting Western scientific research sites asserted that they would not be capable of keeping up the same self-command that their foreign colleagues appeared to exhibit in (they thought) resisting the temptations of food, alcohol, and socializing that was necessary to build good *guanxi*. This was true even for the young well-educated scientists who coveted the imagined truths that this lifestyle supposedly brought. "It was very relaxing, but I would feel too lonely," Li told me. She liked this kind of life for a few months, she explained, but going home every day and cooking for oneself and then going to bed was just not how Chinese did things. Instead, Chinese people "like to go out with friends." And because that was part of their essential nature, Li said, that made it difficult if not impossible for them ever to be quite as "objective" as their foreign counterparts.

The contrast that Li and others drew between the self-restraint of the foreign lifestyle and the indulgence associated with *yingchou* activities carried an undertone of moralized self-critique that was not universally shared. Older, less-educated public health professionals I knew in Tianmai who had been steeped in decades of *guanxi*-based morality agreed that "going out with friends" was an essential part of Chinese identity, but did not see *yingchou* as indulgent at all. While both they and their younger counterparts saw it as hard and exhausting work, these older folks also saw it as noble and highly moral work: drinking together was not an indulgence or a breech in purity but rather an important moral obligation. It may not be *danchun* exactly, but it wasn't what might be implied by *danchun*'s opposite—as somehow debaucherous or uncouth. Instead, the willingness and ability to engage in the more complex entanglements of work and leisure that were demanded of Chinese public health scientists retained a moral valence.

But it was exactly the existence of a line between work and leisure that for Li and many of her young colleagues imbued foreign scientists with a higher morality. A lack of personal moral obligations to others meant that foreign scientists could focus on their obligations to the science. The result would, Li thought, be an uncontaminated form of objectivity and the production of a purer "truth."

Situated Trust

How might this understanding of the multiplicity of truths serve as a Chinese intervention into the understanding of trust and truth in science studies? Many STS scholars have understood trust as something that is produced by subjective people and predicated on relationships (Hardin 1992, 2006; Shapin 1994). Hardin (2006) describes trust as "encapsulated" in reciprocal expectations, and he insists that those expectations are specific. Most people don't normally "trust" something or someone in a general abstract sense; rather, that trust is embedded in a particular situation. According to Hardin, "Trust is a three-part relation: A trusts B to do x. Typically, I trust you to do certain kinds of things. . . . Only a small child, a lover, Abraham speaking to God, or a rabid follower of a charismatic leader might be able to say 'I trust you' without implicit modifier" (1992, 154). Hardin's interpretation represents a sort of inverse of previous Chinese interpretations of the situatedness of truth, in which trust is entirely dependent on the specific relationship context. What matters in the Chinese *guanxi* relationship is precisely that A trusts B, regardless of x. And yet the particularities of the A and the B are very important: trust cannot reign where either A or B is unknown or abstract. As Fei Xiaotong put it in his discussion of Chinese sociality and the moral rules that guide it, "General standards have no utility. The first thing to do is to understand the specific context: Who is the important figure, and what kind of relationships is appropriate with that figure? Only then can one decide the ethical standards to be applied in that context" (1992, 78).

From Hardin's point of view, trust in Western contexts frequently adds a layer of embeddedness (as per Shapin's interpretation). One not only has to figure out the appropriate relationships for a given context, but also the important end product of the relationship. Who is doing what for whom? And yet Hardin's exception of the small child, who trusts just because his experiences in the past have led him to trust and he has known no deception, also reflects a perception that China's public health scientists had of foreign scientists. Foreigners trust like small, innocent children—even when the specific A and B are unknown—and that is both admirable and at least a little bit pitiable. Foreign scientists do not need to know the particularities of a particular B or x, because they have a baseline level of what Yan Yunxiang (2009) calls *social trust*—they trust that others will be good to them or at least benign, regardless of who they are. Hardin explains, "Being an optimistic truster [like the small innocent child] opens up the opportunity of great loss and of great gain, neither of which might be possible without risking trust" (1992, 161). For many of China's new public health professionals, Chinese scientists—being incapable of being optimistic trusters—lose that possibility of gain. That is one reason they worry they will remain hopelessly behind. Without the ability to trust widely and without specificity, they fear they will never be able to reach the gains that their Western counterparts are able to reach.

Conclusion

At the outset of this article I noted that public health science after SARS was supposed to "save" China in part by promoting the reputation and standing of China's public health community throughout the world—thus establishing its members as trustworthy professionals and global citizens. With regard to this promise, the young, well-educated scientists who worked at local CDCs in the Pearl River Delta in the years following the 2003 SARS epidemic were largely disappointed. They had arrived at their new posts full of idealism and ambition, but by the end of the decade they were voicing frustration, boredom, and embarrassment. They were deeply bothered by what everyone knew but no one really cared to acknowl-edge before—that nobody knows what the "real" numbers are; that they are perhaps inherently unknowable; and, most troubling, that all of the numbers they spend all day crunching could very well be made up, or at least fudged. They did not feel at all like trustworthy professionals, and this knowledge had become unbearable to them.

What bothered these young scientists most of all was their feelings of power-lessness with regard to acting on this knowledge. The *lingdao* must still receive his statistics, the money must be spent, the migrants must be counted and vac-cinated. Who has time or ability to change a system that cannot be changed, steeped as it is, my interlocutors felt, in thousands of years of their own, dysfunc-tional "culture"? In this vein, the most commonly repeated phrase I heard from the younger generation of scientists whom I met during the course of my field-work was one of fatalism: *meiyou banfa, dou shi lingdao anpai de* (there's nothing to be done, it's all determined by the leaders). Here self-Orientalism mixed with defeatism to reproduce a system that more and more people felt was inherently flawed. Modern science was contrasted with traditional clientelism to produce an ideal that felt impossible to attain.

Even as defeatism reigned with respect to producing a broader system of trust in which "real truths" could be produced, however, Xu Dan and her colleagues continued to seek for themselves those ideals that they insisted were impossible to attain as a collective. For as fatalistic as they were in considering a possible reform of the *system*, a high sense of ambition reigned when considering a possible reform of the *self*. By doing good research at her public health institution in Tianmai—by producing useful, quality data as best she could within the confines of her imper-fect context—Xu hoped she might one day win a chance to conduct postdoctoral research abroad, and join the world of the dull, the objective, and the "real."

NOTES

This chapter is based on research that I conducted in the pseudonymous city of Tianmai between 2008 and 2014. I conducted participant observation and interviews with over 100 informants at over a dozen government-affiliated public health institutions in Beijing, Guangzhou, and Tianmai. While readers familiar with China might easily discern Tianmai's identity, I use this pseudonym in the interest of providing some basic protections to my informants. All other names used in this chapter are pseudonyms. Some identifying details also have been changed or obscured to protect confidentiality.

1. Note that this was true even for university-based public health projects, which invariably still had to go through government institutions like local CDCs in order to proceed, and therefore still required a great deal of *guanxi* and *yingchou*. While it might seem that government-affiliated scientists would be particularly prone to producing nonobjective science, my informants working at universities complained of very simi-lar problems.

References

Blum, Susan. 2007. *Lies that Bind: Chinese Truth, Other Truths*. New York: Rowman and Littlefield.

Carpenter, Daniel P. 2010. *Reputation and Power: Organizational Image and Pharmaceutical Regulation at the FDA*. Princeton, NJ: Princeton University Press.

Daston, Lorraine, and Peter Galison. 2007. *Objectivity*. New York: Zone Books.

Fei Xiaotong. 1992 [1948]. *From the Soil: The Foundations of Chinese Society*. Translated by G. G. Hamilton and W. Zheng. Berkeley: University of California Press.

Fleck, Ludwik. 1979 [1935]. *The Genesis and Development of a Scientific Fact*. Edited by Thaddeus J. Trenn and Robert K. Merton. Translated by Fred Bradley and Thaddeus J. Trenn. Chicago: University of Chicago Press.

Giddens, Anthony. 1990. *The Consequences of Modernity*. Stanford, CA: Stanford University Press.

Gieryn, Thomas F. 2002. "Three Truth Spots." *Journal of History of the Behavioral Sciences* 38(2): 113–132.

Gitelman, Lisa, and Virginia Jackson. 2013. "Introduction." In *"Raw Data" Is an Oxymoron*, edited by Lisa Gitelman, 1–14. Cambridge, MA: MIT Press.

Hardin, Russell. 1992. "The Street-Level Epistemology of Trust." *Analyse and Kritik* 14: S.152–176.

Hardin, Russell. 2006. *Trust*. Cambridge, MA: Polity Press.

Kleinman, Arthur. 1995. *Writing at the Margins: Discourse between Anthropology and Medicine*. Berkeley: University of California Press.

Latour, Bruno, and Steve Woolgar. 1986 [1979]. *Laboratory Life: The Construction of Scientific Facts*. Princeton: Princeton University Press.

Luhmann, Niklas. 2000. "Familiarity, Confidence, Trust: Problems and Alternatives." In *Trust: Making and Breaking Cooperative Relations*, edited by Diego Gambetta, 94–107. New York: Oxford University Press.

MacPhail, Theresa. 2014. *The Viral Network: A Pathography of the H1N1 Influenza Pandemic*. Ithaca, NY: Cornell University Press.

Mason, Katherine A. 2015. "H1N1 Is Not a Chinese Virus: The Racialization of People and Viruses in Post-SARS China." *Studies in Comparative International Development* 50(4): 500–518.

Mason, Katherine A. 2016a. *Infectious Change: Reinventing Chinese Public Health after an Epidemic*. Stanford, CA: Stanford University Press.

Mason, Katherine A. 2016b. "The Correct Secret: Discretion and Hypertransparency in Chinese Biosecurity." *Focaal: Journal of Global and Historical Anthropology* 75(2016): 45–58.

Shapin, Steven. 1994. *A Social History of Truth: Civility and Science in Seventeenth-Century England*. Chicago: University of Chicago Press.

Yan, Yunxiang. 1996. *The Flow of Gifts: Reciprocity and Social Networks in a Chinese Village*. Stanford, CA: Stanford University Press.

Yan, Yunxiang. 2009. *The Individualization of Chinese Society*. New York: Berg.

Yang, Mayfair M. 1994. *Gifts, Favors, and Banquets: The Art of Social Relationships in China*. Ithaca, NY: Cornell University Press.

CHINA'S ECO-DREAM AND THE MAKING OF INVISIBILITIES IN RURAL-ENVIRONMENTAL RESEARCH

Elizabeth Lord

As we chatted under the plastic greenery hanging from the ceiling of his favorite teahouse, Professor Li argued that environmental researchers had grown cold-hearted. They had been lured by money, power, and status away from conducting transformative research; their hearts were not in it anymore. Li condemned contemporary environmental researchers as unconcerned with the plight of others, tangled in quantitative modelling, and unable to generate breakthroughs. His criticisms were sad rather than harsh, but they also felt urgent. As a retired environmental researcher from the Chinese Academy of Sciences who had seen the field evolve, he implied that something was out of joint (Interview BJ42 2014). While it is true that environmental research has not been immune to the "bourgeoisification" of academia, corruption, and the draining of expertise toward the business fields (Hao 2003), an analysis of the politics of environmental research points to a different interpretation than Li's. Rather than a loss of heart or a fading commitment to meaningful environmental research among individual researchers, this study details *structural* parameters that constrain environmental research and shape what type of science is being produced in contemporary China.

In response to acute environmental problems, the Chinese government has invested massively in environmental research over the last two decades. It has cranked up its environmental rhetoric and implemented ambitious environmental protection policies that stretch across the country. And while the ascent of an environmental era in China is beyond doubt, the parameters that regulate environmental research in the country remain fairly opaque. This chapter starts from the premise that technologies of knowing the environment are molded by broader political and societal contexts. Just as "science" is never singular, but rather an ever-changing product of personal commitments, institutional struggles, and historical legacies (Greenhalgh 2008), environmental research, data, and methods are never dryly environmental. Moreover, spatial differences affect how environmental science touches the ground, especially in a context where the rural-urban divide remains a powerful organizational analytic. This chapter therefore focuses on the production of environmental knowledge as an explicitly political process, one that is in constant conversation with institutional, ideological, and economic forces (Goldman, Nadasdy, and Turner 2011). Examining environmental knowledge production is important because China's green dream reaches the population unevenly and builds on inequalities to realize itself.

This chapter identifies key parameters that regulate rural-environmental research. These include pressures to prioritize economic growth over environmental protection, the commercialization of academia, governmental controls of what is considered acceptable or unacceptable research, as well as limitations on fieldwork access. Such pressures seep into the institutional set-up of environmental research, environmental discourses, research design, the methodologies used, and the data that are ultimately produced around environmental questions. This chapter details how these economic and political parameters bound environmental research in specific ways. By evaluating the epistemological effects of these forces on environmental research, I identify concrete limitations that curtail what environmental researchers are able to study and know about rural-environmental questions; I also find that some researchers are able to circumvent powerful pressures by developing effective methodological alternatives.

This analysis brings to the fore the many *invisibilities* that permeate rural-environmental research in contemporary China. Hyams (2004) warns against assuming that silences, or invisibilities, are mere absences or spaces empty of meaning. When taken seriously, invisibilities, gaps, and silences reveal fault lines. Focusing on these invisibilities and on the "epistemology of silence" provides insights about complexities and tensions that constitute the whole—the visible and the less visible (Joniak-Lüthi 2016, 198). Analytical gaps are generated, for example, when certain questions are not asked because of state restrictions, when the focus of an environmental study is overly economic at the expense of other vectors of analysis, or when fieldwork is abruptly interrupted. These gaps can be seen as methodological failures; they can also be read as meaningful data that tell us more about China's environmental project, how it is framed, and how environmental knowledge production works. Focusing on invisibilities highlights the limits of what is discussed and what is effectively erased from discussions and points to what is outside the realm of solutions—namely, the political—in a process akin to what Ferguson details in *The Anti-Politics Machine* (1990).

This chapter is based primarily on one year of fieldwork in Xi'an conducted in 2013 and 2014. During this time, I interviewed environmental researchers, villagers, officials, and NGO workers in Xi'an, the neighboring Qinling Mountains (Shaanxi), Wuhan, Hong Kong, and Beijing.[1] Interviews with environmental researchers revolved around descriptions of their work and what they saw as opportunities, limits, and pressures that regulate their daily academic activities. Our conversations focused on rural-environmental research, and I inquired about data gathering, methodological decisions, funding opportunities, and publications. We discussed the relationship between policies and research, the effect of slogans on their work, and how they envisioned past and future transformations of environmental research in China. I also carried out participatory observation in Shaanxi's rural areas and at environmental conferences. Attending these conferences and reviewing their abstracts allowed me to witness the performance of environmental research and identify key themes and favored methodologies, as well as some of the silences at work in public environmental events. These conferences served as laboratories for considering what is regarded as credible or authoritative scientific knowledge and what is muted.[2] In this chapter, I draw on insights from five environmental departments across the country (in Zhejiang, Hubei, Beijing, Shaanxi, and Hong Kong), which remain anonymous. In addition, and as I will discuss shortly, my own experience conducting environmental research in rural Zhejiang (2009), Hubei (2010–2011), and Shaanxi (2013–2014) and the methodological challenges I encountered initiated (and made necessary) this reflection on environmental knowledge production and on how specific forces shape what can be known, or not, about rural-environmental questions.

Building the Ecological Civilization across the Rural-Urban Divide

In 2007, President Hu Jintao announced that his government intended to build an "ecological civilization" (*shengtai wenming*), an innovative—if open-ended—formulation that promotes ecological integrity and, some say, social equality (Huang 2010; Xu 2013). In 2012 "ecological construction" was added to China's constitution, an amendment that effectively transformed the core definition of "socialism with Chinese characteristics." While "socialism" previously rested on *economic, political, cultural*, and *social* construction, it now also hinged on *environmental* construction (Lü 2013; State Council of the People's Republic of China 2015). Four decades earlier, in 1978, official national policy shifted from "class struggle" to "economic construction," indicating the onset of opening reforms and initiating the reorganization of all spheres of life (Yin and White 1994). The genealogy of the ecological construction can thus be traced back to the erasure of the analytic of class (Anagnost 1997; Zhang 2002), which is relevant considering that China's ambitious environmental project maps onto a deeply unequal society.

Indeed, China's stark rural-urban divide organizes, and in some sense defines, citizens and landscapes. Notwithstanding sporadic spurts of care, thirty years of rural neglect and city-focused developmental policies have created a divided society in which rural people have become second-class citizens (Chan 2009; Yan 2003). This divide has given rise to, and perpetuates, a rural-to-urban subsidy in the form of exploitable and disposable labor, land for cities to expand onto, a "low-stakes" space to absorb the undesirable "externalities" of production, and now a surplus space that can be mobilized to "green" the nation. Pressing environmental concerns thus rub shoulders with deep socioeconomic ills—a reality that compels us to examine China's current environmental project in connection to inequality and rural questions.

In this chapter, considering that institutionalized inequality shapes China's social fabric, the rural-urban analytic serves as the angle from which to examine environmental research. Drawing on this analytic helps us see how China's ecological dream engages different segments of its population differently. To be clear, I draw on the rural-urban divide as a socioeconomic analytic with the understanding that reality does not squarely fit such a rigid binary: urban poverty is on the rise, many "urbanites" are officially registered as rural residents (through their household registration, or *hukou*), and urban peripheries are often neither truly urban nor rural. Still, the analytic is valuable because it conveys core institutional, discursive, and economic differences that are routinely mobilized by many, including the government, to describe, plan, and organize Chinese society: rural-urban differences are mobilized for development, economic strategies, taxation, and environmental policies. Perhaps most important, the rural-urban analytic is an everyday category used by people and communities to define and contest their own identities and to organize their lives. Last, I draw on this imperfect analytic because we must name the realities we seek to criticize.

Ecological Work and Sacrifice in China's Countryside

Before focusing on environmental research and to provide some background for that discussion, I first offer a brief account of some of the rural-environmental dynamics currently transforming the Chinese countryside. Research in the Qinling

Mountains and in other rural areas confirmed that rural communities are intensively mobilized to enact China's ecological dreams. It also revealed deeply contradictory rural-environmental processes.

First, and perhaps surprisingly, many rural communities are required to "green" the nation, or provide urban areas with clean "natural" resources. In the Qinling Mountains, for example, rural communities are asked to shut down industries, relocate, and plant trees to protect water quality for urbanites as part of the South-to-North Water Diversion Project (*Nanshui beidiao*) (Magee 2011; Webber, Crow-Miller, and Rogers 2017). More specifically, the water that flows down the southern side of the Qinling Mountains accumulates in the Danjiangkou reservoir before being diverted all the way to Beijing and Tianjin as part of the middle route of the South-to-North Water Diversion Project. Somewhat dissonant with everyday rural concerns, signs perched on mountaintops ask people in the Qinling Mountains to "Send clean water to Beijing."

Other environmental initiatives that target rural communities include major reforestation projects, such as the Sloping Land Conversion project (*tuigeng huanlin*), which mobilizes 32 million households across the country to plant trees and control erosion (Gutiérrez Rodríguez et al. 2015; Zhang, Zinda, and Li 2017). The promotion of rural and ecological tourism in all provinces also builds on the logic by which rural communities are expected to create "green" value packaged for urban consumption. The massive scale of these ambitious greening initiatives illustrates how rural communities are aggressively enlisted to build China's ecological civilization.

Paradoxically, and perhaps less surprisingly, many rural areas are also prey to the ruralization of pollution (Zhang 2014). This process is fueled by a series of institutional arrangements that protect cities, thus sending pollution, dirty industries, and waste to rural areas. Stricter environmental regulations and higher fees in cities push polluting industries out, while debt-ridden rural governments pull them in for local investment; the rural thus becomes a "sink" for pollution, following a "spatial fix" logic (Harvey 2003). Urban environments are better monitored than rural ones, and new environmental investments tend to flow toward cities (Xie 2007; Zhou et al. 2008). Moreover, even when environmental monitoring of rural areas does take place, a "dual mechanism of environmental regulation" imposes stricter rules in cities (Xia 2011, 7; Zhao, Zhang, and Fan 2014, 179). This means that the same water could be considered toxic in cities and safe in neighboring rural areas where fewer chemicals are tested (Interview XA13 2014). This monitoring and data gap undermines the most fundamental efforts at documenting rural-environmental realities.

Environmental work on the part of rural communities, together with the ruralization of pollution, illustrates how rural communities are enlisted as "ecological soldiers" for the good of the nation and urbanites. Rural people are mobilized as part of efforts to green China, but they are also asked to bear the harm of pollution—and these processes often take place in the same area, such as the Qinling Mountains. Whether considered separately or as constitutive parts of a coherent national greening project, these rural-environmental dynamics carry the risk of reproducing inequalities. The ingrained belief in China that rural people and places are worth less than urban ones (Anagnost 1997, 2004; May 2011; Yan 2003, 2006) justifies the differential that makes the rural malleable for such environmental transformations, both positive and negative. The state-led environmental project not only unfolds unevenly onto the national landscape, but also builds on inequalities for traction. In light of these rural-environmental dynamics, it appears urgent to examine how environmental researchers negotiate the fraught interplay among rural, economic, and environmental imperatives.

This chapter therefore steps back to interrogate the processes by which rural-environmental research takes place and the frameworks that circumscribe what is known—and not known—about rural-environmental issues. In what follows, I focus on some of the parameters that shape rural-environmental research, namely, (1) the weight and pervasiveness of a pro-growth ideology when dealing with environmental questions, (2) the growing pressures and incentives to commercialize research, and (3) the political and methodological limitations that surround rural fieldwork. This analysis thus provides a snapshot of some of the limitations and "biases" that regulate rural-environmental inquiries. The goal is to examine how a seemingly neutral, "scientific," and positive project such as the greening of the country superimposes itself onto a matrix of inequality, which in turn, shapes environmental knowledge and what is knowable about rural-environmental issues.

Environmental Dreams in an Economic World

In the interviews I conducted for this research, environmental engineers, environmental scientists, environmental legal scholars, and environmental economists provided different perspectives on the impact of slogans on their research, the ease of attracting funding, and on how they envisioned the future of the field. To some extent, their specialties structured how they understood the intricacies of environmental research. However, they all agreed on one point: China's environment would improve as the economy continued to grow. Across the board, interviewees argued that economic development would bring environmental protection. As one of them explained: "Developed countries like Canada have solved their environmental problems. I know because I've seen Niagara Falls' clean waters. This is what China is working toward. We need to grow the economy, and environmental protection will come" (Interview XA18 2014).

The idea that economic growth leads to environmental protection echoes the Environmental Kuznets Curve (EKC), an economic model according to which environmental degradation first increases as the economy grows until gross domestic product reaches a certain point, after which environmental degradation starts to decrease (Clapp and Dauvergne 2005). This widespread understanding serves to justify prioritizing economic growth. According to this model, boosting economic growth is a legitimate means to reduce environmental harm and bring about "protection." The EKC is misleading because it draws on a limited aspect of the Western experience—namely, the fact that certain environmental indexes improved roughly at the same time as the economy grew. It erases the fact that many environmental harms (for example carbon emissions) have continued to increase as economies grew. The logic behind the EKC also erases the importance of laws, institutions, and activism in transforming the environmental landscape, but most importantly, it silences the fact that polluting production and waste were exported en masse outside the borders of these more "developed" and increasingly "green" countries—a shift that was pivotal to their environmental improvement. That all the researchers I interviewed voiced different iterations of the EKC (albeit without naming it) signals a widely shared mindset in which prioritizing the economy does not clash with environmental objectives, but rather is an indirect way to bring about environmental protection.

This mindset can be observed in many official documents. For example, providing guidelines for the then-upcoming eighth five-year-plan, the government's 1991 *Environment Statistical Yearbook* stated that "environmental protection science and technology must turn toward the general policy of economic construction," specifying that research must "deepen coordination between the environment and the economy" and that the *ultimate* objective is to "support national and local economies, the macro-policy of the continuation of a stable society, and the promotion of the scientific foundation" (MEP Yearbook 1991, 170).[3] Ten years later, environmental debates surrounding China's admission to the World Trade Organization (WTO) in 2001 were similarly weighted heavily in favor of celebrating economic possibilities over environmental threats. Abigail Jahiel, a political scientist who interviewed officials and environmental researchers at the time, noted that policy experts at the State Environmental Protection Administration (SEPA) anticipated a decrease in industrial pollution and improvement of rural environments.[4] Jahiel emphasizes that these projections were not surprising, considering that the government curtailed dissent, "embraced neoclassical economics, and staked its future on the promised benefits of WTO membership" (2006, 311). The government had much to gain from boosting positive predictions.

More positive predictions anticipated increased environmental capacity through economic growth, improved environmental technologies, and stricter environmental regulations. Less positive predictions emphasized the detrimental impacts of producing at an exponential scale, the risks of creating a "pollution haven," and the potential harm of a regulatory race to the bottom to generate more globally competitive prices (Jahiel 2006). In light of resistance to the WTO in other countries, particularly with regard to potential threats to labor and environmental protection, the Chinese government anticipated disagreements by instructing academics to "depict a rosy picture for the post-WTO era" (Chen 2009, 30). The overall argument was that trade liberalization, accelerated foreign investments, and economic growth were bound to benefit the environment, and research should reflect this.

Despite this optimism, the more pessimistic predictions of the environmental effects of China's WTO admission ended up materializing. Admission to the WTO brought down trade barriers and further accelerated low-cost production. As a result, the scale of production spiked. By 2004, foreign trade had doubled, and foreign direct investments had increased by 30 percent (Jahiel 2006). International corporations were welcomed and conveniently profited from weaker regulation enforcement. Economic reforms of the 1980s and 1990s made it possible to expand low-cost and large-scale production to the global scale, and this precarious balance was sustained by weak environmental compliance and poor labor protections. While many benefitted lavishly from increased trade and booming production, a World Bank study found that the living standards of China's rural poor decreased since the country was admitted to the WTO, and that environmental harm was significantly worse than had been anticipated (Jahiel 2006). Pressures to shape academic discussions surrounding admission to the WTO illustrate how the nexus of state-market-science can operate, especially in such high-stake cases.

A more concrete example of the state's prioritization of economic over environmental goals can be observed "in action" in written exchanges concerning international reports. For example, comments appended to a World Bank (WB) document titled "China: An Evaluation of World Bank Assistance" (World Bank 2005) show the intricate making of environmental discourses and how the relationship between economic growth and environmental protection is negotiated. Of particular interest is a formal letter written by a representative of the Chinese government to the WB writing team, which suggests corrections to the original report (World Bank 2005, 115–119). Thanks to this useful display of WB transparency, readers can see what Mr. Zou Jiayi, the then–deputy director general of the Ministry of Finance's International Department, disagreed with in the report and what types of changes he recommended on behalf of the Chinese government. Although these few pages of comments appended to a WB report are but a detail in the grand scheme of (environmental) things, they nonetheless illustrate the ideological framework within which environmental researchers in China must operate. Zou's comments can be read as state-sanctioned ideals for the relationship between economic growth and environmental protection.

In this formal letter responding to an earlier version of the WB report, Zou objects to the following statement: "A falling water table in the northern China [*sic*] implies that the rate and pattern of growth which has taken place in that region are unsustainable" (World Bank 2005, 119). This statement refers to a well-known water problem in Northern China and the realization that existing water tables have been overexploited. This has led to highly polluted waters regionally, the emptying of local water tables, and ground subsidence. The original wording blames economic development and a destructive "pattern of growth" for lowering water tables in the North. Yet Zou takes exception to the idea that the current development model has detrimental environmental effects. Instead, and through an artful reversal of cause and effect, Zou suggests the following statement: "The descending of the underground water level in the North of China has restricted seriously the sustainable development of the region" (World Bank 2005, 119). This proposed revision erases the human impact and the effects of development on water tables and instead implies that limited natural resources are to blame for unsustainable development. In this new formulation, the reader could assume that water tables were always depleted and that now they hinder sustainable development. The official and final report states that a "falling water table in northern China has seriously restricted the sustainable development of the region" (World Bank 2005, 3), showing that Zou won this negotiation.

Rationalizing water depletion in the North in this particular way is key to justifying the South-to-North Water Diversion project, mentioned above. In order to legitimize the project, the Chinese government has created an environmental discourse wherein local development, namely of the *Jingjinji* cluster (which includes Beijing, Tianjin, and Hebei) is not to blame. Instead, poor natural "endowment" of water resources in the region requires the diversion of water from elsewhere. Implicitly, this pro-growth discourse makes the diversion of water from the South essential for sustainable development to take place. As Webber, Crow-Miller, and Rogers have documented in more depth, it also presents water diversion as an "apolitical project that conceals the anthropogenic sources of water stress on the North China Plain" (2017, 373). Moreover, this pro-growth environmental discourse erases a more complex and multilayered story in which the depletion of water tables in the North calls for the diversion of water resources more than a thousand kilometers away, the shutting down of rural industries, and the mobilization of entire rural communities to protect water for the benefit of distant and developed urban mega-regions. Sacrifices that might be required from rural communities in the Qinling Mountains to protect water that will flow to Beijing and Tianjin are in this way "justified," but also erased from the story.

While these discursive snapshots do not represent the government's entire strategy toward environmental change, they nonetheless illustrate deeply held assumptions adopted and circulated within government circles regarding the relationship between economic growth and environmental protection, and about which spaces matter. They also send signals to researchers (in this case both Chinese and international) about the state-sanctioned way to tackle environmental and economic frictions.

Of course, the belief that economic growth will ultimately bring environmental protection is not limited to the Chinese context; it is a powerful notion that circulates globally. In light of this, the fact that all my interviewees agreed that economic growth would eventually bring environmental protection can be interpreted in a few ways: these opinions may be personal and the product of their own experience and empirical observation, but most likely they are (also) the product of fairly consistent messaging that hammers in the fact that economic growth will eventually benefit the environment. These ideas circulate through diffuse discursive channels among colleagues, in local and international media, at conferences, and in official government documents. Such powerful emphases erase whole angles of analyses and compel us to ask what environmental stories and analyses have *not* been told over the last decades because researchers had to comply with the premise that economic growth would bring environmental protection. As the next section shows, economic pressures also materialize in the organization of environmental research, a process that further generates epistemological invisibilities.

Making Research Make Money: The Commercialization of Environmental Knowledge

As the previous section showed, economic concerns persistently overshadow environmental ones discursively. Here, I focus on how economic concerns infiltrate the very structure of environmental research, notably through the commercialization of knowledge. Commercialization of research, in this context, includes the many pressures to comply with market logics and, more specifically, to align research designs, data, methodologies, or final "products" with the goal of earning financial compensation or even making a profit. It also includes the increasingly close relationship between academia and the industrial complex. To understand how the commercialization of research shapes environmental questions, this section presents a brief overview of the evolution of the commercialization of research in contemporary China, which began over three decades ago. A foray into Chinese academia's ties to industries in the 1980s and 1990s shows the legislative and political foundations of this process.

The push to bring academia closer to industries was initiated in the early 1980s and consolidated in the 1990s. In 1982, the *yikao mianxiang* guideline instructed practitioners that economic development should "rely on" (*yikao*) science and technology, and that, in turn, science and technology should "turn to" (*mianxiang*) economic growth (Eun, Li, and Wu 2006, 1337). This guideline crystalized the importance of the research-industry partnership and encouraged universities to tailor their research toward economic construction. In 1985, the "Resolution on the Reform of Science and Technology System" reasserted the state's commitment to research-industry partnerships and the commercialization of research (Yin and White 1994, 218). Beyond words and guidelines, the state backed the resolution by decreasing funding to universities (Eun, Li, and Wu 2006).

As universities increasingly needed to fund themselves, this shift fueled a growing number of university-run enterprises, which included publishing houses but also hotels, restaurants, and factories. As "chronic financial deficits" plagued universities, the number of fee-paying students increased (a process initiated in the early 1980s) and boosted the quantity and scale of industrial partnerships (Yin and White 1994, 228). Aching for funds, universities in the 1990s adopted an "ever-increasing emphasis on the practical and utilitarian aspects of knowledge," which was compounded by a brain drain toward the business sector (Yin and White 1994, 229, 232). In 2001, the state set up technology transfer centers specifically to promote commercialization in six universities, and in 2002 a controversial debate about the commercialization of academic knowledge was concluded by the then–minister of education, Zhou Ji, who eventually declared that the official core mission of universities was teaching, research, *and* commercialization (Wu 2010, 208). The Ministry of Education responded by releasing a directive encouraging academia-industry linkages.

Since about 2005, the number of university-run enterprises has started to decrease (Wu 2010; Eun, Li, and Wu 2006),[5] but the commercialization of research takes a variety of other forms. In the environmental departments where I conducted research, *consultancy* was one of the main ways academics could earn external funding. For lack of something more tangible to sell, the rise of environmental consultancy allows some social scientists to commercialize their skills in the same way that pharmaceutical laboratories sell their products or chemists might commercialize their findings. As universities ask departments to self-finance, many environmental professors have developed consultancy by mobilizing their academic resources, including knowledge, expertise, research equipment, student labor, and so on. In all five environmental departments where I conducted interviews, students and professors were deeply engaged in some form of environmental consultancy.[6]

This is important because rural-environmental monitoring, rural pollution control, and questions about environmental justice can hardly be shoe-horned into a market or a profit logic. Indeed, in contrast to research on biotechnology or green energy, questions about the plight of rural communities—caught in the middle of increasing pollution and policies to protect resources for the nation—are at odds with the imperatives of commercialization and profit-making. Moreover, given the extent to which those who commission the research can orient results, as I will discuss below, this work can hardly be considered "scientific"; yet, pressured by the economic context, established and fledgling environmental scientists spend much of their time conducting consultancy work.

Being hosted by environmental departments for year-long periods meant that I could witness daily, and sometimes participate in, specific consultancy projects. In one case, I joined a large team of students to identify and catalogue plant species along roads as part of the city's plan to "green" its highways. In another case, a rural town requested an annual development plan that had to comply with environmental requirements, a project I call the Huatang Annual Plan. The local government sought the services of an environmental planning professor, who had at her disposal a large team of graduate students to carry out the work. This particular professor was academically successful and respected, yet she earned a modest salary and felt pressured to conduct consultancy work to bring in funding for her department. The team of graduate students thus embarked on a lengthy process of designing an environmental and developmental plan, which involved multiple fieldwork trips to study local realities. Once the Huatang Annual Plan was devised, students presented it to local officials for approval, but it was repeatedly rejected and sent back for revisions. Local officials persistently demanded less focus on the environment and more on developing the local economy, and each iteration incrementally diluted what had initially been a respectable *environmental* plan. As graduate students explained to me, the team had to comply with these demands in order for the project to be adopted and for payments to be processed to their professor and department. Most students seemed exhausted and frustrated by the string of revisions that stretched over many months, and some of them seemed disillusioned about the possibility of real environmental change. One laughed the whole thing off, arguing that this outcome was not surprising considering the self-interestedness of local officials.

Indeed, the Huatang Annual Plan took many months and was time-consuming for students and professor alike, leaving little time for their own research. Meanwhile, the not-so-environmental development plan carried academia's seal of approval, with all the authority it confers. More positively, the exercise taught students professional skills such as how to interact with local officials, present work in nonacademic settings, and complete a comprehensive annual plan grounded in the specificities of a locality. It also exposed students to rural-environmental realities, including the fact that many local governments are besieged by debt. Yet, while hands-on consultancy work offers relevant learning experiences, financial reasons overshadow pedagogical ones in fueling the rise of consultancy work taking place in environmental departments.

Moreover, while low salaries create a powerful incentive to commercialize research, many researchers disparage the trend, particularly as it eats away at academic resources, time, and energies. Many interviewees also noted how the rise of consultancy has led to stark income and resource disparities not only between departments, but also among faculty *within* departments (Interview WH11 2014; Wu 2010; Yin and White 1994). As a director of an environmental department explained, stark disparities are being consolidated within his university because certain fields can secure major funding through domestic and international investments while others cannot. For example, his work on forest ecology cannot compete with biotechnology, and the brand-new potato research lab across the street from his older building is a daily reminder of a university-wide funding divide. His research must rely on governmental funding or on consultancy, a financial arrangement that shapes what he conducts research on (Interview WH11 2014). But being caught in this wave of commercialization and consultancy work, and complying with it, does not translate to agreeing with it; in fact, most of the graduate students and professors whom I asked about consultancy expressed unease and resigned aversion to the process.

Fieldwork Interruptions and Data Invisibilities

While the previous sections illustrated how economic logics infiltrate and organize discourses, emphases, and structure of environmental research, I now turn away from economics to consider questions of power and methodology—specifically, the methodological parameters that regulate rural-environmental research. This section focuses on researchers' abilities or inabilities to access rural areas, and how this shapes the type of environmental data that can be gathered. Although accessibility to research sites has markedly increased since the

1980s, both the central government and local officials can still curtail environmental fieldwork, and a complex series of reference letters from higher echelons must typically be obtained to gain access to an area, although these still do not guarantee smooth research. Thus, the question here is what types of invisibilities plague rural-environmental research and what gaps in data are created as a result.

Over the three times I conducted formal environmental research in rural areas of Zhejiang, Hubei, and Shaanxi, my fieldwork was abruptly interrupted for various reasons. In the first two cases, I was working with a team of Chinese researchers to interview villagers on specific environmental policies. Our questions had been revised and tweaked by graduate students and professors who were leading us to these rural sites, where they had connections with officials. In the first instance, we asked a villager about the compensation he received in exchange for reforesting his land. The villager replied that he did not know about the compensation and had not received it. The local official who had been hovering over the interview interrupted to say that this person had little understanding of financial matters, but also that he had been late in paying villagers this year and the year prior. After an awkward back-and-forth between the villager and the official, who concluded he would pay the villager his due, we were made to feel, without being told directly, that we had done enough interviews, and we left a few hours later. We were not officially kicked out, but we withdrew to avoid antagonizing local officials. Our professor's access to the site was contingent on personal relations and social networks that loosely mixed professional and affective bonds, and this embeddedness compelled us to avoid behaviors that could be interpreted as confrontational.

In the second instance, I asked local officials about environmental policy implementation, which included a question about the percentage of state-owned versus collectively owned land in the area. My interview guide had been preapproved by my host department, but the officials seemed struck and offended by the question, and after comparing me to a Japanese spy, they terminated the interview and directed us to conclude fieldwork. It later transpired that my professor had not anticipated the sensitivity of questions around land-use rights in this area. In fact, this site had been selected because environmental policies were successfully implemented. In a third instance, I was introduced to a village and its leaders by an NGO worker. The leaders approved my research and agreed for me to stay in their village to investigate environmental protection policies, but on a further visit, the local police said I had to leave because my papers were not in order. When I insisted, the officers suggested that I might be an undercover tourist guide from Canada because I had visited this village multiple times with different people over the course of the summer.

There is nothing surprising about this, and many researchers I spoke to also experienced abrupt interruptions of their rural fieldwork. And although these political and methodological limitations may be more common and rigid for Western researchers, they also affect Chinese researchers (Cao 2000; interviews in Hong Kong, Xi'an, and Wuhan 2014). It is not only the fact that certain questions are considered politically sensitive that curtails research, the *uncertainty* about what is considered sensitive at any given time generates a chill on its own (Heimer and Thøgersen 2006). As a Chinese environmental researcher working on Poyang Lake explained, the degree of welcome fluctuates in sync with changing political moods, and different environmental topics have varying effects on local officials (Interview WH13 2014). These fluctuations may be related to sensitive local concerns, shifting government priorities, politico-economic anxieties, or simply a fear of an investigative gaze into local affairs.

For example, this researcher used remote sensing to study the effects of sunlight on photosynthesis and algae on Poyang Lake. When he started working on this project around ten years earlier, he was warmly welcomed by local researchers. They treated him to food and provided the institutional and material support he needed to conduct his research. Four years later, their enthusiasm had been replaced by a series of bureaucratic hurdles intended to verify the researcher's identity and the "real" objective of his research. Their role had switched to keeping intrusive research out. Poyang Lake, downstream from the Three Gorges Dam, has been significantly transformed by the building of the dam, which has disrupted the lake's natural seasonal flooding cycle. These changes are central to Poyang Lake's vegetation and ecosystem health. While research on erosion and silt, for example, is supported and encouraged by the central government, research on Poyang Lake is now heavily monitored.

In assessing fieldwork constraints on environmental research, Professor Tang's description of data-gathering is also instructive. Professor Tang is a scholar from Hong Kong who regularly conducts fieldwork on ecotourism on the mainland. Because what he told me resonates with what other researchers noted, I note here Tang's description of the political and methodological hurdles he experienced when researching rural-environmental questions in rural Sichuan. He explains:

> It's really hard to get data. There is no room to negotiate with Chinese officials. Even if you try to design your project to fit their agenda, they will delete your questions. You have to get collaborators and take out the politics out of your data-gathering. The project I was involved in brought a huge amount of donation money. We were forced to cut out all negative questions, or questions that could be answered in the negative.

> I told them, okay, I'll do it this time, but it's the last time. And so I was hired as a consultant. The regional officials particularly don't want you to know what goes on in their locality. They don't want you there. It's fine to do research in a big city, you can use questionnaires. But for us, local officials accompanied interviewers [students] for a while and then told them to stop. There were no political questions in our questionnaires. (Interview HK30 2014)

Most striking in this account is that even though researchers were keenly aware of potential political sensitivities, they miscalculated acceptable limits—a methodological misstep that cost them the rest of their fieldwork. Questions had been prescreened and "no negative" questions meant that whole fields of reality would remain empirically invisible. Local officials facilitated fieldwork to some extent, but they also hovered over students who gathered data from households, ready to use their veto power at any point in the process. Tang also emphasized the power of gatekeepers and the need to establish strong local collaborations in order to conduct smooth fieldwork. Later in our conversation, Tang highlighted the increasingly fraught relationship between research and economics. Here he commented on the economic pressures he experienced while trying to gather data in rural Sichuan:

> When you do research on the mainland, you need to first develop good relations with local officials, then you must show how your research will benefit them. Basically the idea is to find common ground between what you want to know and what they might also want to know. Consultancy work sometimes goes against getting data for research. You need to see where there is a demand for certain data—and usually this demand is for data that will have an economic value. You must persuade them that the data will be valuable to them economically. It's a little bit like a dilemma between what's academic and what's practical. For local officials, academic endeavors don't always make sense. (Interview HK30 2014)

As Tang points out, not only do economic pressures increasingly infiltrate environmental research, but researchers must find common economic denominators and actively tailor research to local needs. This implies a less extractive relationship between researchers and local communities—a laudable objective, but also one shaped by economic objectives that may be narrowly defined by local officials or even by their own personal interests. Furthermore, researchers are pressured to conduct research within topics already of concern to state officials in an approach that builds on local efforts and expertise, but also stifles new lines of inquiry.

In response to these methodological challenges, many of the environmental researchers I spoke to combined formal and informal fieldwork. In a typical formal fieldwork setup, a senior researcher might arrange a meeting with municipal or provincial officials to introduce the research of a junior scholar. If approved, officials might select a village or county where the research will take place and make a phone call to this locality. The selection process is opaque, but researchers tend to be sent to places where environmental policies have been successfully implemented and where the official has connections. These formal arrangements typically provide institutional support, food (often feasts), and housing, but may be limited in time and breadth. Less official routes, or informal fieldwork, offer less institutional support, but in some cases a more thorough access. Ways of conducting informal fieldwork might include knowing someone from a village—a strategy used by both Chinese researchers (Interview WH14 2014) and foreign researchers (Lora-Wainright 2013). Local researchers might also choose to study their own hometown (Gao 1999), and foreign researchers, taking advantage of the need for English teachers in rural schools, might teach English in order to enter a community in which to conduct fieldwork (Interview XA03 2014). Informal fieldwork strategies often rely on friends or family ties to determine site selection, which constitutes an obvious sample bias, but these strategies allow for an in-depth and more personal understanding of the landscape and people. In contrast, formal fieldwork might disproportionately focus on eco-villages, "environmental" parks, or other model areas where environmental policies have been successfully implemented and where the emphasis falls on "happy" environmental stories. While informal fieldwork might be limited in scope, the blind spots of formal fieldwork include sites where rural-environmental problems are considered too vexed and political. Both strategies hold the risk of interruptions, but combining the two approaches may increase the chance of accessing rich and extensive, and in some cases more systematic, data.

Conducting "Safe" Research through Data-Recycling

Due to state restrictions on fieldwork and potential frictions with gatekeepers, researchers constantly develop alternative data-gathering strategies. Professor Li, the somewhat disillusioned researcher mentioned in the opening, shared what he described as a well-known trick to avoid overstepping political lines: use existing data, especially governmental data. The Ministry of Environmental Protection (now the Ministry of Ecology and Environment) publishes more data than one could ever use, and he insisted that environmental researchers, particularly younger ones working through sensitive questions, should stick to the ministry's published material. But he added a caveat: "If it can be published, it's probably not real." And he continued: "If you really want to know what the situation is, you should do your own fieldwork" (Interview BJ42 2014). He was not perplexed by the dissonance of these statements. In fact, that these statements may be simultaneously true and widely known by environmental researchers testifies to the troubled relationships between environmental realities, political forces, and methodological decision-making. Still, Li stuck to his point: use published data. And as I interviewed more environmental researchers, I realized that *recycling* data was a common methodological approach to conduct "safe" research.

A Xi'an-based environmental researcher's dilemma about publishing his student's article on soil pollution illustrates how data-recycling can work (Interview XA13 2014). The student's study mapped a soil pollutant in a rice-producing province, and instead of gathering first-hand data, the student compiled and synthesized data from a wide range of existing studies. The outcome was a worrisome map with precise latitudinal and longitudinal coordinates of locations where the pollutant in question significantly exceeded health standards. More than thirty points of different sizes representing varying levels of toxicity were scattered across the map, next to the names of towns and villages.

While mapping soil toxicity is essential to devise solutions and prevent poisoning, the student's article could not be published in the existing context. This was in 2013–2014, when soil pollution was deemed a "state secret."[7] As the researcher explained, knowing the exact location of toxic soil could lead to social instability: farmers would feel frustrated and powerless, they could rebel or stop farming, and food security could be threatened. The effects of publishing and circulating such precise data on local communities were hard to predict, he thought, but were likely to be negative. At the same time, the study was politically safe to the extent that it was devised from existing data; the graduate student perused all sources available in Chinese and English up to that point (data were gathered before 2013 when the "state secret" label was affixed) and mapped the results. No fieldwork, first-hand research, approvals, or soil samples were necessary. Synthesizing existing data enabled the research, the map, and the precise findings to exist.

But the disturbing visualization of this toxic rice-producing province was disruptive, if not subversive. Nonetheless, the professor felt compelled to get the information his student had compiled out. And so, he pondered aloud, one loophole was to publish the article in English and hope that the insights were picked up by researchers in China as they bounced back through global knowledge networks. Studies published in English might be monitored less closely because of their limited readership in China, and they garner more points on the "bibliometrics" ladder. Another strategy was to conceal the content of the article under a convoluted or jargon-heavy title. Committees that decide on promotion and funding, he explained, are likely to read only the title—an obscure title could conceal more audacious content.

As they debated the next step, it was clear that they had to strike a balance between making the research *invisible* enough that it did not draw too much attention and *visible* enough to be caught by other researchers working on the topic.

Conclusion: Rural-Environmental Blind Spots as Data

The massive windows, high ceilings, and white tiles of Professor Wang's satellite campus office made another thick grey Xi'an day seem bright. As the interview unfolded, Wang described his environmental engineering research and the types of pressures, limits, and opportunities that regulate it (Interview XA13 2014). His work consisted of developing water treatment technologies, but he also reflected on slogans such as the "ecological civilization," funding opportunities, and international collaborations. Wang's lab had accepted a request to test the water from a village bordering Xi'an, free of charge. The village stood outside city limits by just a few kilometers, and its water quality was therefore poorly monitored, yet villagers suspected the water was unfit to drink. Sure enough, his lab found levels of fluoride six times greater than those stipulated by health standards. If the city's boundaries had been drawn with a slightly wider pen, the village's water would have been treated easily and long ago. Villagers hesitated about next steps; they had been drinking toxic water for so long.

As he spoke, Wang scribbled characters connected by arrows to clarify his thoughts. In the middle of the scribbled page, he wrote, in English this time, the word "think." He underlined it once, then drew a large square around it, then another, then retraced the lines multiple times. "See how many times I underlined it?" he asked, pointing to the word. "That's how bad it is. We don't have time to think." As a thoughtful and caring scientist, concerned with rural-urban inequalities and committed to improving the lives and environments of the most disadvantaged, Wang felt caught between pressures to publish, constraints on research topics, and expectations from his field of environmental engineering to commercialize his research. The lack of space to devise real solutions to China's most pressing environmental problems left him dismayed.

It is difficult to categorically identify what we do not know about China's environment because of the political and methodological constraints discussed above, but we can speculate that one casualty is systematic and comprehensive rural-environmental knowledge. It is difficult to combine environmental research in rural areas with questions that are already sensitive, such as property rights or how local governments benefit from fines and taxes levied on polluting industries (see Tilt 2007). The difficulty of obtaining rural archival material (noted by both Western and Chinese researchers) reduces the feasibility of longitudinal studies. Sensitivities around realities like "cancer villages" mean that it is difficult to build cross-national comparative data that would be essential to devise solutions to these problems (Lora-Wainwright 2013, 2017). Similarly, survey responses that are obtained in the presence of a hovering official erode trust in aggregate data gathered through surveys because interviewees give different answers in the presence of officials (Lora-Wainwright 2017). Overall, this chapter illustrates that while some researchers are able to conduct in-depth and rich environmental studies, geographical and thematic invisibilities often creep up and steadily accumulate in the data, thus limiting researchers' ability to uncover and document the complexities and specificities of rural-environmental struggles. Moreover, while central government institutions are able to carry out extensive and large-scale research (e.g., the five-year national soil pollution survey), they also hold the power to withhold information if they see fit (e.g., declaring soil pollution a "state secret").

My analysis of rural-environmental science unfolds at a time when China is putting forth a utopian environmental vision, while being tangled in many intractable environmental problems. The fault line between this environmental dream and the dystopian realities on the ground reaffirms the importance of analyzing processes of environmental knowledge production, if only to better understand the nature of China's environmental project and the various "truths" on which it builds. In the opening of this essay, Professor Li decried faltering morals and rising indifference among researchers (which might well be true for some), but my data point rather to a series of *structural* limitations on research. This analysis highlights how the overarching emphasis on economic growth, commercialization of research, state controls of information, and limited fieldwork access together give rise to manifest invisibilities around rural-environmental questions, and this precisely at a time when they deserve more—not less—attention. These economic, political, and methodological limitations clash with the routine work of researchers, many of whom actively strive to produce transformative and meaningful environmental work in a divided society.

NOTES

1. I conducted more than two dozen interviews with environmental researchers, and all but a few took place in Mandarin. I use pseudonyms and abbreviate interview locations (Beijing, BJ; Hong Kong, HK; Wuhan, WH; Xi'an, XA).

2. The conferences I attended include the Third International Conference on Energy and Environmental Protection (April 2014, Xi'an), the Sixth International Symposium on Larger Asian Rivers, which focused on "Sustainability of Water Environment and Ecology-Biodiversity and Ecological Security at River Basin Scale" (May 2014, Xi'an), and the Socio-Cultural Geography Conference (June 2014, Xi'an). I also attended the International Conference on Philosophy Today: Key Contemporary Theoretical Issues (December 2013, Xi'an), which included presentations on pressing environmental questions, on the idea of science, and on the philosophical underpinnings of social development and modernization.

3. China's *Environment Statistical Yearbooks* (*Zhongguo huanjing nianjian*) are published annually by the Ministry of Environmental Protection and provide official accounts of environmental change. In addition to statistics, they include speeches by top leaders, reports by policymakers, and details on institutions, funding structures, ongoing projects, and so on. The Ministry of Environmental Protection is the most important environmental institution at the national level in terms of scale, reach, capacity to gather information, and ability to initiate national environmental change, and the statistical yearbook provides a rough map of political, institutional, and ideological changes within the ministry.

4. The State Environmental Protection Administration (SEPA) is the predecessor of the Ministry of Environmental Protection, which was established in 2008 and reshuffled in 2018 into the Ministry of Ecology and Environment.

5. This may be because earnings from university-run enterprises have been decreasing in the face of increasing competition by private enterprises, and/or because they are being replaced by less rigid institutional arrangements, such as contractual research, licensing projects, and research and development initiatives (Eun, Li, and Wu 2006; Wu 2010).

6. Consultancy is a big business. In 2006, for example, the total value of environmental consultancy conducted by Chinese firms was estimated at USD 4.1 billion ("Staying Ahead in China" 2009). Consultancy services offered by universities, including environmental departments, are often less costly than those offered by private firms and have thus become increasingly in demand, particularly over the last fifteen years (Interview HK07 2014).

7. In 2004, SEPA (precursor to the Ministry of Environment and Ecology) and the National Administration for the Protection of State Secrets (NAPSS) signed an agreement according to which any information that could potentially "affect social stability" was to be considered a "state secret" (HRIC 2007, 174). More specifically, information that seriously threatens social stability was classified as *highly secret*, while information that affects social stability—but not in a serious way—was classified as *secret*. This latter category also include information that can make China lose face at the international level or "create an unfavorable impression in our country's foreign affairs work" (HRIC 2007, 174). Soil pollution can lead to social instability as well as international and domestic criticisms, and remediation is costly and difficult. The sociopolitical realities surrounding soil pollution make it an ideal candidate for state secrecy.

References

Anagnost, Ann. 1997. *National Past-Times: Narrative, Representation, and Power in Modern China*. Durham: Duke University Press.

Anagnost, Ann. 2004. "The Corporeal Politics of Quality (Suzhi)." *Public Culture* 16 (2): 189–208.

Cao, Jinqing. 2000. *China along the Yellow River: Reflections on Rural Society (Huang hebian de Zhongguo)*. Translated by Nicky Harman and Ruhua Huang. New York: RoutledgeCurzon.

Chan, Kam Wing. 2009. "The Chinese Hukou System at 50." *Eurasian Geography and Economics* 50 (2): 197–221.

Chen, Gang. 2009. *Politics of China's Environmental Protection: Problems and Progress*. Singapore: World Scientific.

Clapp, Jennifer, and Peter Dauvergne. 2005. *Paths to a Green World: The Political Economy of the Global Environment*. Cambridge, MA: MIT Press.

Eun, Jong-Hak, Keun Lee, and Guisheng Wu. 2006. "'University-Run Enterprises' in China: A Theoretical Framework for University–Industry Relationship in Developing Countries and Its Application to China." *Research Policy* 35: 1329–1346.

Ferguson, James. 1990. *The Anti-Politics Machine: "Development," Depoliticization, and Bureaucratic Power in Lesotho*. Cambridge: Cambridge University Press.

Gao, Mobo C. F. 1999, *Gao Village: A Portrait of Rural Life in Modern China*. London: Hurst.

Goldman, Mara J., Paul Nadasdy, and Matthew D. Turner, eds. 2011. *Knowing Nature: Conversations at the Intersection of Political Ecology and Science Studies*. Chicago: University of Chicago Press.

Greenhalgh, Susan. 2008. *Just One Child: Science and Policy in Deng's China*. Berkeley: University of California Press.

Gutiérrez Rodríguez, Lucas, Nick Hogarth, Wen Zhou, Louis Putzel, Chen Xie, and Kun Zhang. 2015. "Socioeconomic and Environmental Effects of China's Conversion of Cropland to Forest Program after 15 Years: A Systematic Review Protocol." *Environmental Evidence* 4 (6): 1–11.

Hao, Zhidong. 2003. *Intellectuals at a Crossroads: The Changing Politics of China's Knowledge Workers*. New York: State University of New York Press.

Harvey, David. 2003. *The New Imperialism*. New York: Oxford University Press.

Heimer, Maria, and Stig Thøgersen. 2006. "Introduction." In *Doing Fieldwork in China*, edited by Maria Heimer and Stig Thøgersen, 1–23. Honolulu: University of Hawai'i Press.

HRIC—*See* Human Rights in China.

Huang, Chengliang. 2010. *Shengtai wenming jianming zhishi duben* [The Clear and Simple Reader on Knowledge about the Ecological Civilization]. Beijing: China Environmental Science Press, 2010.

Human Rights in China. 2007. *State Secrets: China's Legal Labyrinth*. New York: Human Rights in China.

Hyams, Melissa. 2004. "Hearing Girls' Silences: Thoughts on the Politics and Practices of a Feminist Method of Group Discussion." *Gender, Place & Culture* 11 (1): 105–119.

Jahiel, Abigail. 2006. "China, the WTO, and Implications for the Environment." *Environmental Politics* 15 (2): 310–329.

Joniak-Lüthi, Agnieszka. 2016. "Disciplines, Silences and Fieldwork Methodology under Surveillance." *Zeitschrift Für Ethnologie (Journal of Ethnology)* 141: 197–214.

Lora-Wainwright, Anna. 2013. *Fighting for Breath: Living Morally and Dying of Cancer in a Chinese Village.* Honolulu: University of Hawai'i Press.

Lora-Wainwright, Anna. 2017. *Resigned Activism: Living with Pollution in Rural China.* Cambridge, MA: MIT Press.

Lü, Zhongmei. 2013. "Roadmap for the Construction of an Ecological Rule of Law in China." Special Issue: "Ecological Civilization and Beautiful China." *Social Sciences in China* 34 (4): 162–170.

Magee, Darrin. 2011. "Moving the River? China's South–North Water Transfer Project." In *Engineering Earth*, edited by S. D. Brunn, 1499–1514. Dordrecht, NL: Springer Science and Business Media.

May, Shannon. 2011. "Ecological Urbanization: Calculating Value in an Age of Global Climate Change." In *Worlding Cities: Asian Experiments and the Art of Being Global*, edited by Ananya Roy and Aihwa Ong, 98–126. Malden, MA: Wiley-Blackwell.

MEP—*See* Ministry of Environmental Protection.

Ministry of Environmental Protection. 1991. *Zhongguo huanjing nianjian* [China Environmental Yearbook]. Beijing: China Environmental Yearbook Press, 1991.

State Council of the People's Republic of China. 2015. *Zhonggong zhongyang guowuyuan guanyu jiakuai tuijin shengtai wenming jianshe de yijian* [Opinion of the Central Committee of China's Communist Party and the State Council on Further Promoting the Development of Ecological Civilization]. http://www.gov.cn/xinwen/2015-05/05/content_2857363.htm.

"Staying Ahead in China." 2009. *Environment Analyst.* https://environment-analyst.com/26278/staying-ahead-in-china.

Tilt, Bryan. 2007. "The Political Ecology of Pollution Enforcement in China: A Case from Sichuan's Rural Industrial Sector." *China Quarterly* 192: 915–932.

Webber, Michael, Britt Crow-Miller, and Sarah Rogers. 2017. "The South–North Water Transfer Project: Remaking the Geography of China." *Regional Studies* 51 (3): 370–382.

World Bank. 2005. *China: An Evaluation of World Bank Assistance.* Washington, DC: The International Bank for Reconstruction and Development. http://ieg.worldbankgroup.org/sites/default/files/Data/reports/china_cae.pdf.

Wu, Weiping. 2010. "Managing and Incentivizing Research Commercialization in Chinese Universities." *Journal of Technology Transfer* 35 (2): 203–224.

Xia, Lijiang. 2011. *Xinnongcun huanjing baohu zhishi duben* [The Manual of Knowledge of Environmental Protection for the New Countryside]. Beijing: China Labor Protection Publisher.

Xie, Shujuan. 2007. *Zhongguo nongcun huanjing wenti yanjiu: zhidu touxi yu lujing xuanze* [The Study of Rural Environment in China]. Beijing: Social Science Academy Press.

Xu, Haihong. 2013. *Shengtai laodong yu shengtai wenming* [Ecological Labor and Ecological Civilization]. Beijing: People's Press.

Yan, Hairong. 2003. "Neo-Liberal Governmentality and Neo-Humanism: Organizing Suzhi/Value Flow through Labor Recruitment Agencies." *Cultural Anthropology* 18 (4): 493–523.

Yan, Hairong. 2006. "Self-Development of Migrant Women and the Production of Suzhi (Quality) as Surplus Value." In *Everyday Modernity in China*, edited by Madeleine Yue Dong and Joshua L. Goldstein, 227–259. Seattle: University of Washington Press.

Yin, Qiping, and Gordon White. 1994. "The 'Marketisation' of Chinese Higher Education: A Critical Assessment." *Comparative Education* 30 (3): 217–237.

Zhang, Guo. 2014. "Kongzhi chengshi wuran xiang nongcun zhuanyi de falu duice [Legal Measures to Control the Migration of Urban Pollution unto Rural Areas]." *Journal of Henan Institute of Education* (Philosophy and Social Science) 33 (1): 69–70.

Zhang, Xudong. 2002. "The Making of the Post-Tiananmen Intellectual Field: A Critical Overview." In *Whither China: Intellectual Policies in Contemporary China*, edited by Xudong Zhang, 1–78. Durham, NC: Duke University Press.

Zhang, Zhimin, John A. Zinda, and Wenqing Li. 2014. "Forest Transitions in Chinese Villages: Explaining Community-Level Variation under the Returning Forest to Farmland Program." *Land Use Policy* 64: 245–257.

Zhao, Xiaoli, Sufang Zhang, and Chunyang Fan. 2014. "Environmental Externality and Inequality in China: Current Status and Future Choices." *Environmental Pollution* 190: 176–179.

Zhou, Kai, Zhifang, Wang, Xingzhi Peng, and Haichao Luo. 2008. "Chengshi wuran xiang nongcun diqu zhuanyi he kuosan de dongyin jiqi houguo [Motives of Migration and Diffusion of Pollutant from Urban Area to Rural Area and Its Influence on Construction of Harmonious Society of Countryside in China]." *Research on Agricultural Modernization* 29 (4): 471–474.

THE GOOD SCIENTIST AND THE GOOD MULTINATIONAL

Managing the Ethics of Industry-Funded Health Science

Susan Greenhalgh

Since its embrace of reform and opening to global capital in 1978, China has sought to become a wealthy and powerful nation by expanding the role of the market in the governance of social life. An early target for marketization was China's healthcare and public health system. While investing heavily in the advanced biosciences and biotechnologies, the state virtually abandoned healthcare to the private sector, creating a veritable free market in health (Duckett 2011; Hsiao 2014; Huang 2013; Simon and Cao 2009; So and Chu 2016, 119–139). Even as massive changes in lifestyle—including the Westernization of diets and sharp declines in physical activity—left China's people vulnerable to the chronic diseases of modern life, the state also neglected public health (Huang 2013; Li et al. 2016; Mason 2016a). After the disastrous handling of the 2003 SARS epidemic, the state rebuilt the public health network, yet the overwhelming emphasis was on preparing for infectious disease threats (Mason 2016a). Until very recently (around the mid-2010s), chronic diseases—which make up nearly 90 percent of total deaths (with cardiovascular disease, cancers, chronic respiratory diseases, and diabetes leading the way)—remained political and thus scientific backwaters, with little support from an overburdened health ministry (WHO 2018a).

Not just in China, but in countries around the world the adoption of neoliberal development strategies has led to profound changes in the political economy of health and health science. State attention has shifted from public welfare to market creation, corporations have sought profits in "health," and science has been redirected to commercial value creation. Science studies scholars, some worried about the corruption of scientific research and the perversion of the aims of science (Hackett 2014), have turned their attention to the rise of "commercialized" or "neoliberal science," mainly in Euro-America (Abraham and Ballinger 2012; Jones 2009; Lave, Mirowski, and Randalls 2010; Wadmann 2014). If, as science studies suggests, science-making is a normative process in which scientists routinely incorporate normative reflection into their daily practices and ethical values get embedded in the process of science-making itself (Pickersgill 2012), how do scientists schooled to think of science as objective and value-free manage the ethics of making science when industry is a partner? By the ethics of science-making, I mean the moral principles, or values concerning what is right and what is wrong, that govern scientific conduct.

These questions take on some urgency in the Chinese context, where state-market-science/technology are tangled tightly together to form a knot of governing logics, practices, and institutions (see introduction to this volume). A state that cozies up to large firms, a weak, underfunded public health field hungry for global connections, and a loose ethical climate around government-business-science ties combine to form an inviting climate for food and beverage multinationals seeking allies in China's community of chronic disease specialists. Although corporate involvement in China's health science is now extensive, like other aspects of Chinese science it remains little studied. How do Chinese scientists who are concerned about the worsening crises of chronic disease create health science and policy in a context in which a lack of state support leaves them little choice but to look elsewhere for funds, including to multinational corporations that may have money but also profit-making motives? How do they navigate the institutional, political, and cultural terrain to ensure that their scientific practices accord with their own and the state's and wider culture's ethical norms? More generally, what counts as ethical scientific practice in the hyper-marketized context of contemporary China?

Since the early 1990s there has been growing concern in Chinese scientific and official circles about scientific misconduct, a problem that plagues scientific communities everywhere. Starting around 2005, revelations of rampant plagiarism, data falsification, and other forms of problematic behavior prompted the establishment of ethical oversight committees, offices, rules, and regulations by government institutions and universities (Resnick and Zeng 2010). Even as China continues the search for effective means to discourage such behavior, the Western media have been circulating a narrative of Chinese science as especially vulnerable to ethical lapses because of the power of money and the centrality of personal relationships (*guanxi*) in the political culture (Resnick and Zeng 2010).[1] Observed from afar and through a Western ethical lens, China appears to be a site of pervasive scientific fraud and misconduct. Whenever a new practice comes to light—false peer reviews (Sonmez 2015), gene editing of human embryos

(Tatlow 2015; Wee and Chen 2018), or fetal cell transplantation (Song 2017 and chapter 3, this volume)—it tends to be fitted into this overarching story of Chinese scientific corruption. But ethics are situated—that is, conditioned by the political-economic and cultural conditions of possibility—and what looks patently fraudulent to a Western observer might appear deeply ethical to a Chinese scientist, who is making normative decisions in a context altogether different from that faced by Western scientists. Though some Chinese scientists may sometimes step outside of what are seen as internationally accepted norms, their ethical practices must be understood in their own terms and as products of the local context.

A handful of anthropologists has done just that, illuminating how Chinese specialists struggle to do good biomedicine and science in the context of the global judgment of Chinese science as always already unethical. In some cases, Chinese researchers have internalized the judgment, transforming it into an auto-critique of Chinese culture as the source of merely (politically) "correct" science, which can never compete with the "real, objective" science made in the West (see Mason 2016a and chapter 4, this volume, on public health researchers). In others, they have pushed back, offering alternative ethical accounts and creating new standards of evidence for their treatments to address the critiques (see Song 2017, and chapter 3, this volume, on fetal cell transplant surgeons). In both cases, ethics work was central to the process of science-making itself, and the experts' values left a big imprint on the science that got made.

This chapter examines the practice of ethics in the making of Chinese obesity science. A disease in itself and a risk factor for many other diseases, obesity's prevalence is rising rapidly around the world (including in China), leading the World Health Organization (WHO) to label it one of the greatest public health threats of the century.[2] Obesity is also a pressing concern because private industry is intensely interested in the condition. In the absence of safe, effective, long-term treatments that work for most people,[3] companies in the food, fitness, restaurant, and pharmaceutical industries have sought to grow their profits by marketing a widening array of products as promoting weight loss. Meanwhile, companies in the soda and fast food industries, widely seen as major contributors to the obesity epidemic, have sought to protect their profits from core foods and beverages by reshaping the scientific narrative about the causes of and solutions to the epidemic (Nestle 2015; Taubes 2016). Corporate attempts to influence science virtually always create conflicts of interest. In China, despite growing interest in scientific misconduct, less attention has been paid to other aspects of scientific integrity, such as conflicts of interest in the funding of research.

Here I consider the case of a prominent nutritionist and chronic disease specialist, Chen Chunming, who, after decades as a government health researcher and official and facing limited state funding for chronic disease work, in the early 1990s

established an NGO-type organization that took the lead in naming and addressing the obesity epidemic in China. Virtually all the funding for the organization's
obesity work came from foreign corporations, including food and soda companies with compelling interests in the definition and management of the condition. Over time, through the influence they wielded in this organization, large soda
companies quietly reshaped the science and policy of obesity—in ways that served
their interests, not those of China's people. The Coca-Cola Company had the
keenest interests in obesity, and it is our focus here.

Western observers are likely to see this as another instance of Chinese ethical
laxity producing predictably harmful effects. In her conversations and writings,
however, Chen expressed pride in her accomplishments on the obesity problem,
narrating her actions as deeply ethical. In this chapter I ethnographically examine Chen's situated ethics to understand how she achieved that sense of science
well done. The science studies scholar Martyn Pickersgill (2012) has argued that
science is marked by normative uncertainty in which it is often unclear what the
right thing to do is. With formal rules covering only so much, informal ethical
practices rooted in researchers' own sensibilities come to govern much of science.
The way these informal ethics articulate with scientific practice is intrinsically social. The political-economic and cultural contexts not only provide the conditions of possibility for ethical science, but also leave their stamp on the science
that gets made.

In this chapter I explore Chen's situated ethics in the making of obesity science (and chronic disease science more generally) over fifteen years (1999–2014)
to understand how she managed to do what she considered good science with corporate funding.[4] The first section introduces the International Life Sciences Institute (ILSI), the unusual industry-funded scientific nonprofit whose China
branch she led. During these years, ILSI was the lead organization on obesity research and interventions in China and became more influential than even the
Ministry of Health (MOH). The first section tells the story of how the Coca-Cola
Company, working through ILSI, succeeded in making its (medically dubious)
exercise-first solution the predominant approach to obesity, skewing the strategy not just of ILSI, but of the Chinese government. In the second and third sections, I turn to Chen's core ethical practices, tracing their roots in the distinctive
culture and political economy of contemporary China, then showing how they
worked together to protect Chen and ILSI from normative critique. The conclusion underscores the dangers of this approach to science ethics and places responsibility for strengthening Chinese scientific ethics squarely in the hands of the
party-state, whose policies—especially the marketization of public health and its
own neglect of ethical issues—bear ultimate responsibility for the ethical practices of individual Chinese scientists, including those we meet here.

This chapter is part of a larger project on the role of the global pharmaceutical, food, and soda industries in the making and managing of China's obesity epidemic. Research conducted during 2013–2016 involved ethnographic fieldwork, documentary research, and web-based research. In late 2013 I spent three months in Beijing conducting fieldwork on the post-1990 history of obesity science and policy in China, focusing on the applied or public health branch of the field. At its core was a set of lengthy, semistructured interviews with twenty-five individuals, including most of the top researchers in the applied branch of the field. Fifteen of the interviews were with individuals personally involved with ILSI's obesity work. Although industry funding of health research was not an especially delicate issue in China, my informants were aware of its sensitivity in the United States and so were somewhat uncomfortable talking about it. Ideally, a study of science ethics would emerge from conversations and observations carried out over months or years. That was not possible here; because of the difficulty of the topic, it was clear that an effort to conduct follow-up interviews at a later date would not be welcome. In 2015 and 2016 I interviewed ten leading obesity specialists in the United States and Western Europe for international perspectives on the management of the Chinese and global obesity epidemics.

This chapter draws on the interviews with ILSI's leaders, as well as interview data with the wider set of informants. The discussions with the Chinese experts were full of tacit ethics-talk—talk about right and wrong—and layered reflections on ethical aspects of science. I focus on shared ethical stances, interpreting and contextualizing them to create a picture of the practical ethics that guided ILSI's leaders. The chapter also draws on the semiannual newsletter of ILSI's China branch. The newsletters contained news items on all the activities sponsored or organized by ILSI, providing an important historical record of all of ILSI's obesity projects over the years. Because ILSI was the predominant organization working on Chinese obesity and it collaborated with the major health organizations in China on all important activities (China's MOH and Center for Disease Control and Prevention [CDC], as well as major UN agencies), we can safely assume that the ILSI newsletters cover the great majority of obesity-related activities held in the country.

ILSI-Global and ILSI-China: Corporate-Funded Science

I first met Chen Chunming (1925–2018) in her bright, crowded, book-stacked office on the ninth floor of the CDC building on Nanwei Road in southeast Beijing. Still formidable at age eighty-eight, she was energetic, articulate, and animated about the achievements of ILSI-Focal Point in China (below, simply ILSI-China). Chen chose to start her story in 1999, when obesity landed on ILSI's agenda, but she had an illustrious background that was critical to her later success ("Editorial in Chief (sic), Prof. Chunming Chen" 2018). Chen studied agricultural chemistry and nutritional science at National Central University (now Nanjing University), graduating in 1947. For the next thirty-five years she conducted nutrition research at the Chinese Academy of Medical Sciences (variously named) before becoming director of the MOH's Department of Health and Disease Prevention (1982–1984). In 1983 she became the founding head of the Chinese Academy of Preventive Medicine (later renamed the China CDC), where she served as director until 1992. In 1993 she formally left government service to establish the nongovernmental health science and policy entity ILSI-China, where she remained until her death, initially as director (1993–2004) and then as senior advisor (2004–2018). Her close collaborator, the prominent food toxicologist, food safety expert, and academician (*yuanshi*, Chinese Academy of Engineering) Dr. Chen Junshi, served as deputy director in the earlier period and has been ILSI-China's director since 2004. I also interviewed Chen Junshi, who was equally engaged and knowledgeable, and his perspectives will be an important part of the story told below. (Below, I refer to Chen Chunming as simply Chen and Chen Junshi as JS Chen.) What then is ILSI?

ILSI-Global: How Industry Funds "Quality Science" in the United States

Established in Washington, D.C., in 1978, the International Life Sciences Institute describes itself as a global, nonprofit scientific organization in which scientists from industry, government, and academia collaborate to generate scientific knowledge that serves human health and environmental sustainability (ILSI website). Its founding president, Dr. Alex Malaspina (1931–), was concurrently vice president of the Coca-Cola Company (ILSI president 1978–2001). ILSI is funded by hundreds of member companies, mostly in the food, beverage, chemical, and pharmaceutical industries. It works on a handful of health and safety issues, including, since 1999, obesity. Headquartered in D.C., ILSI has seventeen branches around the world. Each branch agrees to comply with the policies of ILSI's board of trustees in return for charters that allow it to use the ILSI name and receive other benefits.

ILSI is incorporated as a nonprofit, tax-exempt "charitable, scientific, and educational" corporation under U.S. law, a status that requires that its activities be primarily for public benefit (which it claims to do by providing good science), and that it not lobby or make policy recommendations. To maintain this privileged status, ILSI must work constantly to protect the integrity of its science. To avoid potential conflicts of interest or corporate bias of its science, ILSI reportedly maintains a balance of scientific perspectives among state, science, and industry representatives on its board of trustees and scientific committees. It also has a code of ethics that mandates good science and forbids lobbying and attempts to influence legislation (ILSI website).

The unusual role of industry means that ILSI is involved in a distinctive kind of science-making. Its science is not necessarily "bad," but it has special characteristics. I call this the ILSI model of science-making, and it has three main aspects: most of the funding comes from industry; companies have a major voice in agenda-setting; and the organization has an ethic of good science (described below) that is fundamental to its continued operation. This structure is replicated in the branches, with country-specific adaptations that reflect local cultures and political economies.

ILSI-China: Industry-Funded Science, Fitted to China's Culture and Political Economy

The seeds of ILSI-China were sown soon after the birth of ILSI-Global. In 1978 during a visit to Beijing, Malaspina sought out Chen Chunming and invited her to cooperate in health work. In 1993 Chen believed the time had come to create a branch of ILSI in China. In our conversation twenty years later she explained the appeal (IF3).[5] Funds for public health work were extremely tight. Government oversight was burdensome. Establishing a hybrid scientific research organization would allow her to seek funding from a wide range of nonstate sources outside of China. Being better connected with the world would enable her and her colleagues to learn about the advanced ideas and solutions emerging from Western public health science, which could be used to address neglected health problems at home. And even from this nongovernmental organizational setting, she could retain influence with the state because she had been a high health official, and in China, that status remains informally in effect into retirement. (In serving both academic and official advisory roles, ILSI-China was similar to other think tanks in the 2000s and early 2010s; see Zhu 2011.)

In its promotional materials ILSI-China describes itself as a nongovernmental academic institution whose mission is to bridge government, academia, and industry and provide the most current scientific information for policy decisions in the areas of nutrition, especially obesity, food safety, and, since the mid-2000s, chronic disease prevention and control (ILSI-China 2013a, 1). Though its precise legal form remains unclear, ILSI is not a Chinese branch of an international NGO (they were not permitted at the time). Interviews suggest that ILSI is unique, its founding authorized only because of Chen's high political status and reputed connections high up in the central government (IF2; IF16). As Chen explained, ILSI in China is a "special unit within the ILSI family" (IF3). To mark itself as something other than a branch of an international NGO, its founders called their organization "ILSI-Focal Point in China" and referred to its funders as "supporting companies" rather than "member companies" (IF3).[6]

The Chinese variant of the ILSI model gives companies a large say in selecting which research topics and projects the organization will work on. Each year ILSI-China's leaders set standard levels of support, and companies decide how much to give. ILSI then asks the supporting companies to recommend activities. ILSI makes the actual decision on which events to organize. Although Chen maintained that no companies are more important than others, in practice those providing more funds are more influential. An important mechanism for gaining influence is project-specific funding. If ILSI-China requires additional funding for a particular project, Chen explained, it approaches the companies for targeted funds. My own study of some projects receiving special donations suggests that companies providing project-specific funds tend to have a say in how the projects are carried out or, in the case of a conference, are given a place on the program.

A highly unusual feature of ILSI-China, one that distinguishes it from ILSI-Global, is the absence of a board of directors. Chen did not mention that she had no board; instead, she emphasized the exceptional freedom that she (that is, ILSI-China) enjoyed to seek funding from anywhere and to use it as she thought best. Although ILSI-China is affiliated with and physically located within the China CDC (the successor, in 2002, to the Chinese Academy of Preventive Medicine), the CDC has no control over its activities. Instead, she stressed, ILSI "can decide what to do on its own; it is very independent" (IF3). These arrangements clearly reflect China's political culture, in which relations are hierarchical and authority is seen as being rightly centered in (virtuous, capable) leaders such as Chen. (Below I introduce a key element of this culture, Confucian virtue ethics, and show how it shaped Chen's and others' decisions.)

Based on the supporting companies' priorities, as filtered through Chen's (and JS Chen's) sense of what activities would be most useful, ILSI-China put together conferences, task forces, and research activities and invited experts, mostly from China-CDC and Chinese universities, to lend their expertise (ILSI-China 2013a, 2). In many cases, and especially with large-scale activities, ILSI-China co-organized or co-sponsored the event with the leading health organizations in China (in particular, China's MOH and CDC, UNICEF, and WHO). The co-sponsorship gave the events the weight of an official activity while offering participants, including those from industry, personal access to officials. What distinguished ILSI-China from other Chinese entities, Chen promised its corporate supporters, was its focus on major public health issues and its ability to translate science into public policy through a framework she called research-policy-action (IF3; ILSI-China 2013a, 2). Chen's de facto political status as a high government official and her wide-ranging *guanxi*, built up over decades, were vital parts of ILSI-China's extraordinary success. In this and other ways, ILSI-China clearly reflected the political-economic conditions of possibility in China.

In 1999, when ILSI-China, following ILSI-Global, turned its attention to obesity, the organization had seventeen supporting companies. By 2004, that number had risen to twenty-four, and by 2014 it had climbed to forty. The great majority were foreign (mostly Euro-American) multinationals in the food, beverage, and restaurant industries and included some of the world's best-known giants (Coca-Cola, PepsiCo, Nestle, Danone, Hershey's, Mars, McDonalds, and Yum!, owner of Pizza Hut and KFC, among others). Given what these companies sell, one would expect them to have strong interests in the obesity issue.

Making Obesity Science and Policy: Coca-Cola and the Exercise-First Solution to the Epidemic

From 1999, ILSI-China took the lead in addressing the obesity issue in China. Between 1999 and 2003, under Chen's direction and with funds provided by a multinational pharmaceutical company, China's public health community worked quickly to gather relevant data, define obesity as a Chinese disease (with China-specific body-mass index [BMI] cutoffs), and prepare the official guidelines for its prevention and treatment (MOH 2003; this story is told in Greenhalgh 2016). ILSI's foundational work on obesity made it the go-to organization on the obesity question.

In 2004 the focus shifted to the development of public health strategies for obesity control and prevention. In the mid-2000s, multinational food companies were actively mobilized to become partners in this fight (ICN 2005a, 2005b) (the full story can be found in Greenhalgh 2016).[7] China's move was prompted by the WHO's Global Strategy on Diet, Physical Activity, and Health, which assigned companies responsibility in the fight against chronic disease (WHO 2004). Yet China took the corporate role to another level. Representatives of the food industry responded enthusiastically, seeing not just profit opportunities in "obesity prevention" and "healthy lifestyles," but also a chance to be part of the solution to (not the cause of) the obesity epidemic and to stay in the good graces of a party-state that could target Western companies in anticorruption campaigns at any time. By repeatedly stressing the corporate responsibility theme and by creating concrete mechanisms to enable company participation, Chen help insert them directly into the nation's core strategy to combat obesity and chronic diseases more generally.

By around 2006 the ILSI model had become the official mode of scientific policymaking in the area of chronic disease, and ILSI itself, with its corporate agenda-setting, was at the center of the process. In an environment in which the health ministry had little interest in and few resources for chronic disease work and ILSI was the most active and effective actor in the area, what came to count as chronic disease science and policy was in essence whatever ILSI's supporting companies wanted to fund. Of all the companies keen to be seen as fighting obesity, the most visible, in the view of my expert-informants, was the Coca-Cola Company, a regular member of ILSI-China's supporting-company team (IF1; IF4; IF5).

In the very early 2000s, the Atlanta-based Coca-Cola Company (and other soda giants) developed a strong interest in obesity, whose uninterrupted rise was seen as posing a grave threat to the company's profitability (Nestle 2015, 102–104). Intent on getting on the right side of the issue, Coca-Cola declared that it would be part of the solution and that it would make the fight against obesity one of its global sustainability commitments. In the United States it put in place an aggressive, multipronged science strategy to protect its profits. The most important prong was to shift the blame for obesity by emphasizing that the problem is not a poor diet, it is a lack of exercise—a claim few obesity specialists accept (IF1; on the larger corporate strategy, see Aaron and Siegel 2017; Nestle 2015). In China, the company seems to have had a similar science and health strategy, promoting it in multiple venues (most famously, the 2008 Beijing Olympics, which it sponsored). ILSI-China facilitated the company's efforts by providing mechanisms by which companies could quietly influence the topics of research and conferences, have a say in who was invited, and shape health policies and interventions.

From 2004 to 2014, Coke and its China division, Coke-China, became very active in ILSI- and government-sponsored activities to combat obesity and related chronic diseases. Other companies (especially Nestle, the Swiss food and beverage giant) were active too, but none as much as Coke. While the efforts were wide ranging, in scientific conferences and forums, most of the company's energies were directed at spreading the message that physical activity is the key to weight control. Of the activities I have identified, few urged dietary change, and none recommended restricting sugary beverages, both of which are seen by global health leaders as essential parts of an anti-obesity strategy (WHO 2015).

The company also funded public health initiatives that put its message into practice. The earliest and most visible was Happy 10 Minutes, a program to encourage exercise among schoolchildren. Developed by an ILSI unit headed by Malaspina and brought to China by ILSI-China in 2004, Happy 10 became a standard feature of the MOH's national Healthy Lifestyles Campaign (IF5; ICN 2005c, 2005d, 2006a).

To bolster the scientific grounds for the exercise-based approach, Coke also put money into strengthening China's cadre of exercise and sports medicine specialists. In 2011 Coke-China, working with ILSI-China, launched a three-year scholarship program for young professionals to gain short-term training abroad in physical activity and health (ICN 2012a). Coke-affiliated scientists were also highly visible in scientific conferences. In late 2013, Coke was one of four corporate supporters of a major international conference on obesity control and prevention in China. Coke-connected speakers promoting exercise as the main solution to the epidemic dominated the top slots on the program (ILSI-China 2013b). (The Coke connections were not made public.)

In the 2010s, the most important Coke-sponsored project has been Exercise is Medicine (EIM, founded in 2007), a global partnership that encourages health care providers to prescribe exercise as medical treatment. The Coca-Cola Company was EIM's founding corporate partner. Launched in 2012 by ILSI-China and the American College of Sports Medicine, the EIM China Program focuses on training clinicians to incorporate EIM into their practices (ICN 2012b). Through the training and the many related media activities, the health halo surrounding the exercise cure was spread from public health to clinical medicine and to the public at large.

The data suggest that, through its funding of such activities, Coke was able to exert quiet influence on ILSI's and, in turn, China's approach to obesity. Between 2004 and 2014, an ever-larger share of ILSI-China's obesity activities focused on exercise (rather than diet), and in 2010–2014 the majority of activities were exercise-related (Greenhalgh 2019). Though dietary changes were not ignored, China's official policies to combat obesity and related conditions also came to be skewed toward physical activity. Widely recommended dietary policies—especially taxation of sugary drinks, and restriction of food advertising to children—were missing. Though it is impossible to measure precisely, the company's outsized role in the field clearly tilted China's approach toward an emphasis on activity over dietary restriction as the main answer to the obesity epidemic.

Ethical Boundary Work: "Science, Not Business"

If Chen took the lead in allowing multinational companies to influence obesity science- and policymaking, how did she frame the ethics of her actions? There can be no doubt that both she and JS Chen thought that ILSI's working arrangements were not only not problematic—they were beyond ethical reproach. In the helter-skelter environment of ILSI's China—where science ethics are underdeveloped, the state's embrace of big business may suggest that anything goes, relationships are everything, and things are in constant flux—science-making is marked by extraordinary normative uncertainty. With formal ethical guidelines covering only a few practices, many aspects of science-making are informally governed by ethical practices emerging from researchers' own sensibilities. Individual sensibilities are, of course, heavily shaped by the wider culture. Of all the ethical values infusing Chinese culture, the cluster most influential here was what is known as *Confucian virtue ethics* (applied to China in Mason 2016a, 116; see also Wong 2017). As used here, Confucian virtue ethics is a paternalistic, relational ethics in which relations are hierarchical, leaders are expected to be virtuous and to govern ethically, and others, recognizing their leader's moral rectitude, are supposed to reciprocate with loyalty and obedience.

In our conversation and in her many talks and writings, Chen described a loose assemblage of working ethical norms and practices around this issue, which had two parts. In the first, described in the following section, she hewed closely to formal norms established by the Chinese government and by ILSI-Global. Issues that fell outside their scope would be managed according to informal norms guided by her own culturally shaped ethical sensibility. Potentially quite contentious issues—such as funding arrangements—were relegated to a zone of silence, in which her decisions would not be subject to review and debate.

Formal Ethics: Following the Rules

Creating sound science and scientific policy advice was critical to ILSI-China's identity and its legitimacy. Scientists everywhere try to protect the purity of their science by engaging in rhetorical boundary work that distinguishes their science from politics, religion, business, and other supposedly impure domains (Gieryn 1999). ILSI-China's leaders engaged in endless boundary work aimed at cordoning off their science from the commerce that funded it. The most important strategy was to rigorously enforce the Chinese government's rule of "no product endorsement" in ILSI-China-sponsored activities. Supporting companies were not allowed to promote corporate logos or products at ILSI events or to mention their ILSI connection in advertising. ILSI's leaders expressed strong confidence that their firm adherence to this rule fully protected the organization's science. Such views were repeated by many informants who had been active participants in ILSI projects over the years. ILSI leaders also referenced a rule in ILSI-Global's code of ethics, by which no company could influence the science or benefit financially from supporting the organization (IF3). ILSI's leaders sought to reassure me that, unlike in the United States, corporate influence on science was not an issue in China because the rules effectively protected the integrity of the science.

Despite the confident explanations provided in our discussions, tensions around these issues were apparent. Sometimes the code did not quite fit the reality, producing awkward moments in our conversations. For example, I asked JS Chen why the Swiss giant Roche Pharmaceuticals, maker of the anti-obesity drug Orlistat, had generously funded the early work on BMI cutoffs (IF8). Chen first replied: "All ILSI members support all ILSI projects. There is no relation between the support of a particular project and a company's product development or marketing. This is in the ILSI bylaws." (Here he was describing the standard annual payments.) But then, perhaps recalling more of the details, he added: "The Roche case of support for BMI research is an exceptional case." Perhaps uncomfortable with this explanation, which seemed to skirt the ILSI-Global rule, he added that Roche sold not only the anti-obesity drug, but also other products, such as vitamins and minerals. He seemed to be suggesting that the company's interests were much bigger than obesity, and so its funding was not necessarily tied to its hopes for the obesity drug. In this case—and it was a major one, for Roche's funding enabled ILSI to own the obesity issue and get it on the state's agenda—it was difficult to make the rule fit.

In other cases the rules did not work because they were flexibly interpreted by other actors. I witnessed an awkward moment at the 2013 obesity conference in which the chief scientific officer of Coca-Cola, herself a central figure in ILSI-Global, gave a speech that blatantly promoted the company's low- and no-calorie beverage options, violating ILSI-China's cardinal rule of no product endorsement. Clearly, there were limits to ILSI-China's ability to restrict the commercial speech of major supporting companies. Yet there was little its leaders could do except hope the moment passed quickly, and then erase it from history when writing the column about the conference for the newsletter (see ICN 2013b).

Seeking to clarify the larger politics for me, JS Chen explained: "The government has never had any problems with . . . ILSI's corporate support. It trusts that ILSI leaders will use the support well. One cannot say that the government *encourages* corporate support, nor does it *support* it"; he also noted, pointedly, that "there is no criticism of research support by companies in China" (IF8). Others seconded that view, saying that this is just business as usual, unremarkable and unremarked. What he was telling me was that the state was the final arbiter of right and wrong, correct and incorrect. As long as it did not ban the practice of corporate funding of research, scientists could take advantage of the ambiguity by faithfully adhering to the rule the government had laid down and then accepting company funds but doing so quietly. As long as they honored this formal rule, ILSI researchers could feel safe from government criticism and assured that their practices met an officially acceptable ethical bar. And by not talking about the practice among themselves—that is, by surrounding the issue with a strategic silence and refraining from debating it among themselves or formulating shared professional norms—they created a zone of ethical secrecy in which individuals could make decisions in accordance with their own largely unarticulated code of professional ethics. What was crucial was to stay within the state discourse; that, I believe, is why so many of my expert-informants were adamant that there was a rule guiding these matters, that it was critically important, and that they always followed it to the letter.

Sensitive to potential charges of commercial influence, ILSI's leaders publicly emphasized the scientific character of ILSI-China's activities. In a second sort of boundary work, they included the word "scientific" (*kexue*) in the names of their conferences to distinguish them from the many profit-oriented, company-sponsored meetings held in China (IF8). ILSI's many media forums were dedicated to "instilling scientific knowledge and attitudes" in journalists (ICN 2006b). Activities and materials for the general public—educational pamphlets, health classes, BMI calculators, and so on—invariably stressed the scientific nature of the information (ICN 2002). The political weight of the word "science," coupled with the intensely scientistic culture, made it a powerful tool in their kit of boundary-setting practices. Drawing on their reputation as esteemed scientists, Chen and JS Chen were able to place ILSI's activities beyond question simply by attaching the label "science" to them. Put another way, their status as famous scientists helped protect ILSI's science from ethical scrutiny.

Practical or Informal Ethics: Zones of Silence

There was much the formal rules did not cover, however, and here informal discourses and practices—practical ethics—took over. Although I was unable to view these in operation, it is useful to see which issues fell into that zone of informal ethics. One such beyond-the-boundary issue was what the companies got in return for their contributions. In response to my question, Chen explained that they gained access to scientific evidence. When she didn't supply an example, I suggested that presumably she meant evidence they could use in marketing or other business practices. (She neither assented nor dissented.) Asked whether companies ever attempted to influence the science, she replied impatiently, "No, they don't. They are used to these arrangements, they know there will not be any commercial benefit for them." In simply reiterating the formal norms (no company benefit, science is untouched), and rejecting the possibility that the science/business boundary might sometimes be porous, she effectively relegated all questions that did not fit the standard story about the separation of business from science to a zone of silence or unspeakability. For issues in that zone, she could follow her own (culturally informed) ethical sensibility free from interrogation and accountability.

If supporting companies were not supposed to benefit, the whole matter of corporate funding of ILSI—how, how much, for what—raised a hornets' nest of potentially ethically messy questions. Chen seems to have dealt with them by placing them all in that zone of silence and secrecy, where she would be protected from nosy questions by the workings of Confucian virtue ethics in which she was the morally upright leader, owed deference and respect. In my discussions with scientists involved in ILSI activities over the years, I discovered that few knew much if anything about ILSI's funding arrangements. No one seemed to know where the money for the projects they worked on came from or how it was managed. Most suggested that I ask Chen herself. Even staff who organized ILSI conferences and processed the reimbursements "only know how the money was spent, not how it was gotten" (IF35). Some researchers with long experience in the nutrition field were aware that ILSI-China was supported by large food companies, including Coke, but had no information on the amount provided or how the funds were used (IF2; IF12). The organization's written materials were only slightly more illuminating. ILSI-Global's annual reports and ILSI-China's newsletters listed the names of the supporting companies, but no details about how much funding each provided and for what. The articles in ILSI-China's newsletters described each activity in some detail but rarely mentioned the source of its funding. Just as the local health researchers whom Mason (2016b) studied maintained strategic secrecy about certain disease-incidence data to cloak awkward realities (such as data fabrication), for Chen, keeping funding details secret served a number of useful ends, freeing her from dissenting opinions, increasing her operational flexibility, and perhaps in the 2010s offering her some protection from anticorruption campaigns.

The way ethics articulate with scientific practice is inherently social (Pickersgill 2012), and the larger social context—in this case, China's hierarchical political and scientific culture—was crucial. If Chen's status as a virtuous paternalistic leader earned her the loyalty and deference of her followers, so too did her elevated location in the nation's political and scientific hierarchies. My conversations with ILSI experts suggest that not only did Chen not reveal funding details, but others evidently did not feel the need to know them. Nor did they feel they had the right to ask. As head of ILSI, Chen was responsible for obtaining the funding and so had the right to decide how it would be used. But there was more to it than that. In China's hierarchical political system, what followers should and should not know is determined by the leader. One relatively senior scientist based in a top university described Chen as his "teacher" (laoshi), adding that "I did not dare ask, it was awkward to ask" (*bugan wen, buhao wen*) (IF4). The term "teacher" carries various meanings, and it is not clear which one this senior scientist was using. Conventionally, to call someone a teacher is to signal respect and trust, and from the context of the conversation I believe this is what he meant to convey. Yet, as a CDC insider told me, researchers and staff at the CDC call virtually everyone in the organization "teacher" (as in "Chen Laoshi"), stripping the term of its sense of exceptional respect. Nonetheless, Chen was held in high esteem by many of those I spoke to. Though in general people in this hierarchical system take care to display high esteem for their leaders, my informants cited many reasons they held Chen in high regard. People respected her not only because of her political position ("she was the former leader of CDC"), but also because of her energy ("young people could not keep up") and, most of all, because of her scientific expertise ("she was a real expert in the field of basic nutrition") (IF35). As an eminent scientist, a high government official, and the head of an influential nonprofit, Chen was a person of indisputable status, and so people seem to have trusted her to do the right thing. Even if they did not trust her, they most likely feared what might happen if they questioned her, for in China's political culture, to challenge a superior is to challenge the whole system and risk one's future career within it.[8] In China's hierarchical political culture it was neither necessary nor appropriate nor safe to ask how things got done.

Public Narratives about the Good Scientist and Corporation

In a second set of ethics practices, Chen created a public narrative about herself as a good scientist who was global-minded, dedicated to the nation's health, and selfless. Intertwined with this (indeed, co-constructed with it) was a narrative about the good company that, not coincidentally, was global, contributed to the nation's health, and received little in return. If the boundary-drawing practices described in the last section worked to place ILSI's activities on the right side of the ethical line, these notions of the good scientist and company served as overarching narratives that structured the larger discourse, establishing what could and what should not be said about ILSI and its supporters.

The Good Scientist

In our conversation and in her extensive writing, Chen conveyed a clear image of the good scientist she strived to be—and to be seen as being. Three qualities stood out. All reflected the values of the larger culture, especially patriotism and self-sacrifice for the collective good. Not surprisingly, all have been documented in other scientific contexts in China. First, Chen sought to help nudge China out of the backward slot in global science by bringing the best of international science to the local public health community. The ambition to rid Chinese science of its backward label was an ever-present theme in Chen's writings and speeches. As Chen narrated the story of ILSI's founding, for example, the appeal of working with Malaspina and ILSI-Global was not the ample funding but the opportunity it presented to overcome the knowledge gap that was keeping China backward (IF3). Virtually all the conferences she organized through ILSI were aimed at "presenting the latest international science" to the Chinese public health community. Her larger aim was to boost China into the ranks of countries producing internationally well-regarded science, a goal that pervades the sciences in China (e.g., Mason 2016a; Song 2017; Zhang 2012).

Chen's good scientist was also a patriot who served her country by helping identify and solve problems it was facing, or would soon face, and create solutions that met global standards. Her use of funds from multinational food and drug companies was but a means to this larger end. For example, Chen acknowledged that Roche had paid for the foundational research on obesity, but insisted there was no ethical problem because the company did not benefit and because it made a critical contribution to China by enabling ILSI get the issue on its public health agenda (IF3). In Chen's assessment of her work with ILSI, we can hear clearly the notion of science discussed in the volume's introduction: as a practical tool to benefit the Chinese nation. For Chen, as for most Chinese researchers, the good scientist was not one who sought the truth, but one who advanced scientific knowledge for the sake of China's society and nation as a whole.

Finally, for Chen, as for many other Chinese professionals, the good scientist was a selfless scientist who dedicated herself to her country with no expectation of material benefit. She used company funds to pay ILSI's office staff, but, she told me, neither she nor JS Chen accepted a salary. Stories of Chen's exceptional selflessness and public generosity circulated in the public health community. In one account, Chen used her own private funds to establish ILSI-China: "I heard that actually an American supported Chen with much money for her own research, several tens of thousands of dollars," said one interviewee. "But she used the money intended for her own research to build ILSI-Focal Point in China. Others said she was a good person because she gave money . . . to create a nongovernmental organization that the government was able to support [i.e., to authorize]" (IF4). (I was unable to confirm this account but suspect the rumored support from "an American" was the funding provided by ILSI's supporting companies, many of which were American.) In our conversation, Chen described how ILSI had created many of China's obesity policies and quietly gotten them endorsed, only to see them issued in the ministry's name alone. She remarked: "The Ministry of Health has good words to say about ILSI. We're happy that we do good things for the people and the government accepts them. ILSI does not [care about] claiming credit" (IF3).

The Good Multinational

A closely related image was that of the good corporation that contributed generously to China's chronic disease work with little expectation of return beyond scientific evidence. The two images not only had similar features, but were co-constituted, for the good scientist could not perform good works without the help of the good company (and vice versa). Reflecting the general ethos of the reform era—in which economic development and opening to the global market were celebrated as the keys to China's prosperity and success—the good company was almost always foreign. Chen put a distinctive spin on this bias toward foreign corporations. She relayed how the heads of several Chinese companies, invited to support ILSI, had demanded to know what they would receive in exchange. When she replied "scientific evidence," they reportedly expressed disbelief, to which she commented, disdainfully: "Chinese companies don't understand the value of science" (IF3).

I did not ask Chen specifically about the Coca-Cola Company, but in ILSI's newsletters, news items about Coke's contributions to the obesity problem presented Coke in terms fitting the good-company image. For example, stories about Happy 10 Minutes invariably focused on how it improved the nation's health by preventing and controlling weight-related diseases among schoolchildren (ICN 2005c, 2005d). News items about Coke's physical activity scholarships emphasized their contribution to upgrading and globalizing China's sport science. By cultivating "advanced knowledge and global visions for the long-term development of physical activity," one article said, the scholarships helped "enhanc[e] China's overall research [capability] in related fields" (ICN 2013a, 15).

ILSI-affiliated researchers shared this image of the good company, many citing Coca-Cola as a shining example. Reflecting the pervasive narrative about China's backwardness relative to the advanced West, some spoke admiringly of Coca-Cola's work to combat obesity. One highlighted Coke's clever ideas and catchy public health slogans. She was impressed that the company donated money to social welfare causes and that those gifts were mandated by Coca-Cola international (IF7). To these researchers, large Western corporations appeared as sources of not only badly needed financing, technology, and knowhow, but also a much-appreciated ability to get things done. The contrast was with their own government, which was seen as too often corrupt, untrustworthy, and inept. On the whole, large foreign firms were viewed as trustworthy and not corrupt and, given the dearth of state funding, essential contributors to China's public health work.

Marginalized Truths, Unspoken Possibilities

These positive images (or narratives) worked in two ways to structure the wider ethical discourse around ILSI and its corporate supporters. First, they made plain what constituted good science, a good scientist, and a good company; developments that fit the narratives were to be stressed and lauded. Second, observations and interpretations that did not fit were accorded lower priority or not mentioned at all. Things that challenged the narratives were to remain unsaid, except in private.

What kinds of things could not be, and were not, publicly said? One dangerous view, one that a handful of informants held but dared not openly express, was that foreign companies often *do* benefit financially from funding chronic disease work and that funding by interested corporations inevitably taints the science. The few who shared this view, all outside the ILSI circle, expressed their views cautiously and obliquely, and asked anxiously that their remarks remain anonymous. A second unspeakable (and unspoken) truth was that, far from being completely selfless, Chen enjoyed substantial professional benefits from her association with ILSI: extensive foreign travel, membership on important international scientific committees and boards of trustees, and rising prestige at home as ILSI's corporate support and influence surged. No one in China mentioned this to me (though surely they were aware of these perks), but several international obesity experts did. A third discordant thought was the possibility that in their obesity work, what Chen and ILSI were bringing to China was not the most advanced international science, but rather science that was biased in favor of particular, company-friendly solutions. Such biasing could easily have occurred—without ILSI-China's awareness—because Chen's *guanxi* connections tied her most closely to ILSI- and Coke-affiliated international scientists. She may well not have known how controversial their exercise-first ideas were in the international public health community. No one mentioned this possibility to me, but my empirical research shows that the science ILSI-China imported clearly reflected the interests of the soda industry.

In this way, what some outside China might see as Coke's troubling intervention in Chinese obesity science and policy got re-narrated as part of Chen's and ILSI-China's good science and selfless patriotism. For my informants, what was important about Coke's science and projects was not that Coke funded them, but that they were foreign in origin and that (in their perception) they benefitted China. Only a handful of my informants dared to think (out loud) outside this box, and then only in private.

Good Science and the Dangers of Business as Usual

What then counts as good science in China's public health work? It is really quite simple. For Chen and the wider community of researchers, science was ethical if it followed extant formal ethical rules, it helped China solve its problems and become a global power, and it did not benefit the scientist personally. What is notable—and notably Chinese—is how much is left unsaid. Unlike in the West, where the tendency is to create ever more formal rules to anticipate every ethical problem, in this Chinese case it seems that the more left unsaid, the better. In a well-honed if unarticulated strategy, ILSI's leaders seem to have relegated everything that did not fit the rules to a zone of silence, while affiliated scientists, trusting their leader (or feeling some mixture of loyalty, gratitude, and/or fear of the consequences of raising concerns), left ethics alone. For ILSI-China's leaders, relying on silence and secrecy was not just politically useful, enabling them to act quickly without having to justify their actions to the state; it was probably politically necessary, offering protection from the vagaries of an often capricious state with contradictory agendas on science, industry, and ethics. For ILSI's leaders, the zone of silence was a zone of relative freedom and of safety.

Western journalists certainly go too far in framing "unethical China" as the immutable foil to the "ethical West." Yet the approach to science ethics I have documented is in fact full of hazards—for the scientists and for China's people. With so many things rendered unsayable and so many alternative truths marginalized, important questions became unaskable. In a world of huge, powerful, global firms pursuing profits with little regard for human health (Freudenberg 2014), the ethical approach ILSI developed over the years seems to have kept it from seeing the possibility that a favorite American company—the generous one with the friendly vice president, the advanced technology, and the colorful public health slogans—might be working quietly and systematically behind the scenes to shape China's science and policy to its own advantage, to the detriment of the health of China's people. Despite its extraordinarily capable leaders, in the end ILSI-China was too trusting of foreign firms, it relegated too many issues to the zone of silence, and its leaders were too aloof to benefit from the insights of the wider public health community.

The critique of Chinese scientific ethics does, then, contain an element of truth, but it does not justify harsh judgment against the scientists, for it fails to acknowledge the contextual nature of scientific ethics and the lack of options available to researchers in areas with little financial support from the state. The context was decisive. Indeed, each of the limitations just mentioned can be traced back to key elements of the context. The excess trust in firms like Coca-Cola was rooted in the unbending belief at the time in China's scientific backwardness and the West's superiority. (This is rapidly changing; in the late 2010s there is growing confidence in China's scientific and technological prowess.) The excessive silence was a product of the state's own silence about acceptable ties between state and business, politics and economics, money and power. And the leaders' insulation from followers is rooted in the influence of Confucian value ethics and the hierarchical nature of the political culture. If ethics everywhere are contextually shaped, in China the biggest contextual force is the party-state. Ultimately, the ethics ILSI devised were a product of the state's marketization of public health, which left chronic disease experts little choice but to look to foreign firms for funding.

If the Xi Jinping administration (2012–) is concerned about ethical problems in Chinese health science—and it should be, for the sake of China's reputation as well as its health—it needs to devote more attention to ethical issues and create clear rules to guide scientists through the ethical thicket of making science in an era of full-on marketization. That is exactly what my most candid informant suggested, and so I want to close with his thoughts:

> Not many complain about company influence on scientific results. Some people worry, yes, there are some, but their voice is very small. China is unlike the U.S., where it's very sensitive. Risks—definitely there will be. In the area of management, [things are not good], China has not caught up to the U.S. There are no written regulations. There is no conflict-of-interest concept. Chinese are relatively casual (*suiyi*) about this.
>
> The requirement about just not advertising a product—that is too low a standard! Maybe you can suggest some policies for the Chinese government, because we are too weak (*cha*) in the area of corporate influence. I hope our leaders can hear this criticism, and then can [develop] some management that's stricter. Slowly, with modernization, government officials are going abroad to interview [and see how things are done]. Under Xi Jinping, things are a little better now. (IF17; ellipses omitted)

NOTES

The author thanks Jiuheng He, Katherine Mason, Li Zhang, and two reviewers for the press for their trenchant comments on an earlier version of this chapter.

1. On the influence of *guanxi* on public health science, see Mason 2016a.

2. Today some 39 percent of adults worldwide are overweight and 13 percent are obese (WHO 2018b). In absolute numbers, China is the fattest country in the world (NCD Risk Factor Collaboration 2016). Although the prevalence of adult overweight and obesity remains well below that in the United States (42 percent compared to 69 percent), obesity is rising much more rapidly there (Mi et al. 2014).

3. The exception is bariatric surgery, whose success rate is fairly high but whose invasive nature and high cost make it unappealing and/or infeasible as a treatment for most people.

4. On the fraught and contested nature of "good science" in China's public health work, see Mason 2016a, 2016b, and this volume.

5. "IF3" refers to the third interview in my Interview File; the same format is used below.

6. The term *supporting companies* is meant to indicate a more distant connection than *member companies*. ILSI-China's supporting companies could join and leave from year to year, for example.

7. The ILSI-China Newsletter File (ICN) is a document constructed by the author that contains all news items used in the research. The full name of the publication from which the items are drawn is *ILSI-Focal Point in China Newsletter.*

8. I thank Jiuheng He for an illuminating discussion of the internal politics of China's political culture and especially how it plays out in the China CDC.

References

Aaron, Daniel G., and Michael B. Siegel. 2017. "Sponsorship of National Health Organizations by Two Major Soda Companies." *American Journal of Preventive Medicine* 52(1): 20–30.

Abraham, John, and Rachel Ballinger. 2012. "The Neoliberal Regulatory State, Industry Interests, and the Ideological Penetration of Scientific Knowledge: Deconstructing the Definition of Carcinogens in Pharmaceuticals." *Science, Technology, and Human Values* 37(5): 443–477.

Duckett, Jane. 2011. *The Chinese State's Retreat from Health: Policy and the Politics of Retrenchment.* London: Routledge.

"Editorial in Chief (sic), Prof. Chunming Chen." N.d. *Biomedical and Environmental Sciences.* N.d. http://www.besjournal.com/aboutBES/eic/.

Freudenberg, Nicholas. 2014. *Lethal but Legal: Corporations, Consumption, and Protecting Public Health.* Oxford, UK: Oxford University Press.

Gieryn, Thomas F. 1999. *Cultural Boundaries of Science: Credibility on the Line.* Chicago: University of Chicago Press.

Greenhalgh, Susan. 2016. "Neoliberal Science, Chinese-Style: Making and Managing 'The Obesity Epidemic.'" *Social Studies of Science* 46(4): 485–510.

Greenhalgh, Susan. 2019. "Soda Industry Influence on Obesity Science and Policy in China." *Journal of Public Health Policy* 40(1): 5–16. https://link.springer.com /article/10.1057%2Fs41271-018-00158-x.

Hackett, Edward J. 2014. "Academic Capitalism." *Science, Technology, and Human Values* 39(5): 635–638.

Hsiao, William C. 2014. "Correcting Past Health Policy Mistakes." *Daedalus* 143(2): 53–68.

Huang, Yanzhong. 2013. *Governing Health in Contemporary China*. London: Routledge.

ICN—*See* ILSI-China Newsletter File.

ILSI website. N.d. https://ilsi.org/.

ILSI-China. 2013a. *ILSI-Focal Point in China: Two Decades of Achievements and Looking to the Future*. Beijing: ILSI Focal Point in China.

ILSI-China. 2013b. 2013. Conference on Obesity Control and Prevention in China, Theme: Appropriate Technology and Tools for Weight Control, Beijing, December 12–13 [bound copy of program and PowerPoint slides].

ILSI-China Newsletter File. 2002. "Active Public Education on Obesity in China Promoted."

ILSI-China Newsletter File. 2005a. "Round-Table Meeting on Strategy for NCD Control Convened by ILSI FP-China."

ILSI-China Newsletter File. 2005b. "Food Industry Round Table Meeting on 'Strategy on Diet, Physical Activity and Health' Jointly Organized by MOH and China CDC."

ILSI-China Newsletter File. 2005c. "'Happy Ten Minutes' Well Recognized by Students, Parents, School and Government Officials."

ILSI-China Newsletter File. 2005d. "'Happy 10' Project Successfully Tested in Beijing."

ILSI-China Newsletter File. 2006a. "'Happy 10' Project are [sic] Expanding in China."

ILSI-China Newsletter File. 2006b. "Media Forum on Obesity and Related Diseases Control in China."

ILSI-China Newsletter File. 2012a. "The First Batch of ILSI China-Coca-Cola Training Fund Scholarship Winners Completed Training."

ILSI-China Newsletter File. 2012b. "ILSI Focal Point in China Continue [sic] the Promotion of EIM in China."

ILSI-China Newsletter File. 2013a. "ILSI FP-China-Coca-Cola Food Safety and Physical Activity Scholarship Winner Meeting."

ILSI-China Newsletter File. 2013b. "2013 Conference on Obesity Control and Prevention in China."

Jones, Mark Peter. 2009. "Entrepreneurial Science: The Rules of the Game." *Social Studies of Science* 39(6): 821–851.

Lave, Rebecca, Philip Mirowski, and Samuel Randalls. 2010. "Introduction: STS and Neoliberal Science." *Social Studies of Science* 40(5): 659–675.

Li, Yanping, Dong D. Wang, Sylvia H. Ley, Annie Green Howard, Yuna He, Yuan Lu, Goodarz Danaei, et al. 2016. "Potential Impact of Time Trend of Life-Style Factors on Cardiovascular Disease Burden in China." *Journal of the American College of Cardiology* 68(8): 818–833.

Mason, Katherine A. 2016a. *Infectious Change: Reinventing Chinese Public Health after an Epidemic*. Stanford, CA: Stanford University Press.

Mason, Katherine A. 2016b. "The Correct Secret: Discretion and Hypertransparency in Chinese Biosecurity." *Focaal: Journal of Global and Historical Anthropology* 2016(75): 45–58.

Mi, Ying-Jun, Bing Zhang, Hui-Jun Wang, Jing Yan, Wei Han, Jing Zhao, Dian-Wu Liu, et al. 2015. "Prevalence and Secular Trends in Obesity among Chinese Adults, 1991–2011." *American Journal of Preventive Medicine* 49(5): 661–669.

Ministry of Health, Department of Disease Control. 2003. *Zhongguo chengren chaozhong he feipangbing yufang kongzhi zhinan* [*Guidelines for Prevention and Control of Overweight and Obesity in Chinese Adults*]. Beijing: Ministry of Health.

MOH—*See* Ministry of Health.

NCD—*See* Noncommunicable Diseases.

Nestle, Marion. 2015. *Soda Politics: Taking on Big Soda (and Winning)*. Oxford, UK: Oxford University Press.

Noncommunicable Diseases Risk Factor Collaboration. 2016. "Trends in Adult Body-Mass Index in 200 Countries from 1975 to 2014: A Pooled Analysis of 1,698 Population-based Measurement Studies with 19.2 Million Participants." *Lancet* 387(10026): 1377–1396.

Pickersgill, Martyn. 2012. "The Co-Production of Science, Ethics, and Emotion." *Science, Technology, and Human Values* 37(6): 579–603.

Resnick, David, and Weiqin Zeng. 2010. "Research Integrity in China: Problems and Prospects." *Developing World Bioethics* 10(3): 164–171.

Simon, Denis Fred, and Cong Cao. 2009. *China's Emerging Technological Edge: Assessing the Role of High-End Talent*. Cambridge, UK: Cambridge University Press.

So, Alvin Y., and Yin-Wah Chu. 2016. *The Global Rise of China*. Cambridge, UK: Polity.

Song, Priscilla. 2017. *Biomedical Odysseys: Fetal Cell Experiments from Cyberspace to China*. Princeton, NJ: Princeton University Press.

Sonmez, Felicia. 2015. "China's Scientific Research at 'Turning Point,' Study Says." *Wall Street Journal*, November 26. https://blogs.wsj.com/chinarealtime/2015/11/26/chinas-scientific-research-at-turning-point-study-says/.

Tatlow, Didi Kirsten. 2015. "A Scientific Ethical Divide between China and West." *New York Times*, June 20. https://www.nytimes.com/2015/06/30/science/a-scientific-ethical-divide-between-china-and-west.html?mcubz=0&_r=0.

Taubes, Gary. 2016. *The Case against Sugar*. New York: Alfred A. Knopf.

Wadmann, Sarah. 2014. "Physician-Industry Collaboration: Conflicts of Interest and the Imputation of Motive." *Social Studies of Science* 44(4): 531–554.

Wee, Sui-lee, and Elsie Chen. 2018. "In China, Sacrificing Ethics for Scientific Glory." *New York Times*, December 1. https://www.nytimes.com/2018/11/30/world/asia/gene-editing-babies-china.html.

WHO—*See* World Health Organization.

Wong, David. 2017. "Chinese Ethics." In *The Stanford Encyclopedia of Philosophy*, edited by Edward N. Zalta. https://plato.stanford.edu/archives/spr2017/entries/ethics-chinese/.

World Health Organization. 2004. *Global Strategy on Diet, Physical Activity and Health*. Geneva: World Health Organization.

World Health Organization. 2015. *Fiscal Policies for Diet and Prevention of Noncommunicable Diseases*. Geneva: World Health Organization. http://apps.who.int/iris/bitstream/10665/250131/1/9789241511247-eng.pdf?ua=1.

World Health Organization. 2018a. *Noncommunicable Diseases (NCD), Country Profiles*. Geneva: World Health Organization. www.who.int/nmh/countries/chn_en.pdf.

World Health Organization. 2018b. "Obesity and Overweight, Fact Sheet." June. http://www.who.int/mediacentre/factsheets/fs311/en/.

Zhang, Joy Yueyue. 2012. *The Cosmopolitanization of Science: Stem Cell Governance in China*. New York: Palgrave Macmillan.

Zhu, Xufeng. 2011. "Government Advisors or Public Advocates? Roles of Think Tanks in China from the Perspective of Regional Variations." *China Quarterly* 207: 668–686.

THE BLACK SOLDIER FLY

An Indigenous Innovation for Waste
Management in Guangzhou

Amy Zhang

In 2013, Dr. Wu, an entomologist, introduced me to his pilot experimental project to use the black soldier fly (*Hermetia illucens*) to speed up the treatment of organic waste.[1] In his makeshift workshop on the edge of Guangzhou, I watched as Dr. Wu and the caretakers he employed attempted to devise a strategy to breed fly larvae. Trays of fly larvae ready to devour a mixture of organic waste—wet vegetable leaves, fish bones, and other kitchen scraps—were stacked on shelves along one side of the workshop. When the larvae were ready to undergo metamorphosis, they would be transferred to a makeshift garden behind the workshop where caretakers watched and waited as the grown flies mate, lay eggs, and die during the last week of their life-cycle.

A winged insect that originated in the Americas, the black soldier fly (BSF) can now be found in most temperate regions of the world. In the last twenty years, scientists around the world have experimented with using the fly as a form of biotechnology. The adult fly devours organic waste, and its larvae (about 3–19 mm) are marketed as an animal feed rich in protein. Researchers especially emphasize the so-called voracious appetite of the flies to quickly devour all sorts of municipal food waste including slaughterhouse waste and even animal manure.[2] While most BSF projects remain limited in scope, Dr. Wu hoped to devise a method for raising flies that would allow the BSF to be integrated into urban neighborhoods as a part of the municipality's organic waste processing technology.[3]

Dr. Wu's attempts to create a more sustainable method of waste treatment emerged at a moment when the problem of environmental degradation garnered the attention of both the state and civil society. Over the last ten years, leaders in

China have focused on shifting the country's image from the "world's factory floor" and one of its biggest polluters to a global economic power, with a central concern with the environment a critical component and measure of development (Zhang 2017). During the 2008 Beijing Olympic Games, reports of environmental degradation in China dominated international headlines (Macur 2007). Since then, Xi Jinping's administration has enforced domestic emission standards and invested heavily in green technologies. In Xi's China, when it comes to environmental protection, science and technology (S&T) have become critical instruments to produce "translatable results that will help solve enduring economic and industrial problems" (Greenhalgh, introduction to this volume).

Wu's project captures two central concerns in Xi's vision for the rejuvenation of the nation: a concern with innovation as a core aspect of national development and a commitment to tackle pollution and other threats to the environment and health. Scholars of technological innovation in the post-reform period tend to focus on top-down efforts to promote technological innovation, particularly in the sectors of digital technology, telecommunication, and auto manufacturing (Economy 2018). However, Chinese scientists in a broad array of fields have also been concerned with coming up with novel, local solutions and technologies to address the challenges created by rapid development.

Against this backdrop and as part of a search for a sustainable technology for the treatment of organic waste, Chinese scientists proposed using fly larvae as a waste technology. In response to a local problem and buttressed by a history of Chinese entomology, Dr. Wu's BSF project exemplifies a bottom-up approach to "indigenous innovation" in the development of S&T. Until recently, China relied largely on imported technology to achieve scientific and technological modernity. The state, however, has shown increasing support for experiments in local forms of innovation aimed at addressing particularly Chinese problems. In terms of municipal waste management, state policy endorses the construction of centralized technological facilities—primarily waste-to-energy (WTE) incinerators—to treat household waste. However, a high percentage of organic waste in the municipal waste stream makes burning difficult and raises questions about the suitability of WTE incineration as a technology to treat Chinese waste.

I was first introduced to the BSF during my dissertation fieldwork from 2012 to 2013, when I was studying schemes to modernize Guangzhou's urban waste management infrastructure. Over the course of eighteen months of research, I conducted participant observation and interviews with a range of urban citizens—waste planners and local officials, anti-incineration activists, informal scrap collectors, sanitation workers, environmental NGOs, local scientists and entrepreneurs—all of whom were engaged in debates and practices concerned with urban municipal waste management. I became aware of Dr. Wu's project in late 2012 while working with a local NGO investigating a range of technological methods for the treatment of organic waste. I was fascinated by the unconventional approach and began ethnographic research with Dr. Wu in his field laboratory. I returned to Dr. Wu's lab in the summers of 2014 and 2018. During this period, I also interviewed other entomologists working in Guangzhou, Europe, and the United States to more broadly understand the state of BSF research.

In this chapter, I follow Dr. Wu as he works to devise a solution for organic waste treatment using insects. Examining the tension between China's urban development and the sustainable treatment of organic waste, I argue that China's project to institute a green modernity increasingly shows a preference for scientific solutions that address local conditions. This preference is in stark contrast to previous policies and approaches under which Chinese cities, for example, pushed for the adoption of imported waste management technologies and, in the case of organic waste, expelled the animals that served as a de facto waste management system. In the last ten years, the quest for sustainable development combined with a sensitivity to the discrepancy between imported waste technologies and the specific material make-up of Chinese waste have led scientists to experiment with the use of a species of insect as a biotechnology. Dr. Wu's work constitutes a unique example of an "indigenous innovation" (*zizhu chuangxin*) insofar as it is a response to local conditions, it emerges out of a local research practice, and it constitutes a shift away from a reliance on foreign technological transfer. I highlight the longer historical traditions and practices that buttress the development of the technology. China has a specific history of using insects as a tool and resource and has also focused on biological pest control in domestic entomological research. As a waste management technology, the BSF project creates a new use for insects while simultaneously generating increased interspecies dependencies between insects and humans.

Defining "Innovations" in Science and Technology

The concept of innovation (*chuangxin*) refers to a broad guiding principle in Chinese state discourse that mobilizes S&T in service of the goals of national development. According to Benjamin Elman, since the late-Qing period "imperial reformers, early Republicans, Nationalist party cadres and Chinese Communists have all made science and technology a top priority" in the domain of national development (Elman 2006, 1). China's approach to S&T has historically been shaped in one way or another by an imagined lack of modernity in comparison to the West. The notion that the Qing dynasty, particularly the Qianlong emperor, categorically rejected Western technology has been discredited, as scholars have shown a long and sustained interest among Chinese elites in European science (Elman 2006). However, particularly after the Taiping Rebellion, and especially during the Self-Strengthening period (1861–1895), Chinese elites came to understand S&T as more than simply the perpetuation of a tradition of natural studies but as a field and pursuit explicitly linked to state power (Elman 2006, 7). In the late Qing period innovation was thus conceived of in terms of the nation's capacity to turn away from imperial orthodoxy and to embrace modernity. State investments in technical schools, shipyards, and translation bureaus all facilitated a shift away from Chinese tradition toward a wholesale adoption of Western science.

After the late imperial period, China pursued scientific and technological innovation through foreign influence. After the perceived failures of the 1898 education reforms, Chinese students were sent to Western universities to bring modern science back to China to accelerate national development. The Republican elites tasked with modernizing the nation state saw S&T as key to reforming Chinese education, institutions, and practices (Shen 2014). During the late Qing and early Republican period, Japan was an important mediator of science; terms such as *science* (*kexue*) and *technological systems*—including *urban sanitation* (*weisheng*)—entered China through Japan (Kirby 2000; Rogaski 2004).

In the 1950s, the Communist leadership looked to the Soviet Union as a model for scientific research, and during the Maoist period science became explicitly guided by political imperatives (Wei and Brock 2013). After China's official break with the Soviet Union in 1961, China pursued its own program of socialist science that emphasized participation by the masses in the production of scientific knowledge and sought to combine Western and indigenous ways of knowing (Fan 2012). Mao's phrase that science "walks on two legs"—the masses and the specialists—prompted the spread of a patchwork of experimental projects that involved mass participation and aimed to combine indigenous knowledge with expertise in a diverse set of fields from agricultural science to seismology (Fan 2012; Schmalzer 2016).

In the post-reform period, Deng Xiaoping nominally restored authority to experts in the realm of official technology policy (Andreas 2009), and the state turned to large, centralized projects at the expense of local experimentation. At the same time, however, China's emergence as a center of manufacturing also encouraged forms of technological innovation from below (Lindtner 2014). The term *shanzhai*—best understood as "pirated" or "counterfeit"—was first used by scholars in the 1950s to describe the manufacturing practices of small-scale, family-run factories in Hong Kong, which reproduced cheap quality household items (Lindtner, Greenspan, and Li 2015). In the post-reform period *shanzhai* production of digital hardware such as mobile phones (with features designed specifically to suit local usage patterns) exemplified a unique indigenous manufacturing condition and culture in China (Zhu and Shi 2010). The proliferation of *shanzhai* production speaks to the development of an unofficial "maker" culture, one that operates outside of official S&T institutions but nevertheless emerged out of state efforts to build a cluster economy (Zhu and Shi 2010).[4]

After more than thirty years of various approaches, S&T innovation is now a central component of China's strategy of national development. This emphasis is most clearly expressed in the 2012 "Science and Technology Roadmap," which lays out a strategy to make China a "global powerhouse in innovation by 2050" (Wei 2016). The concept of S&T innovation is closely tied to China's economic ascent and efforts to restructure the economy toward higher-value products. "The Medium- and Long-Range Plan for the Development of Science and Technology" (MLP) a 2006 policy document, specifies technological innovation as the path to an "overall well-off society" (*quanmian xiaokang shehui*) (Cao, Suttmeier, and Simon 2006). In 2016, China announced a national R&D program aimed at fostering innovation through investment in clean energy, big data, and computational research. In 2018, China surpassed the United States as the leading publisher of scientific articles, a sign of China's growing influence in scientific research and development (Berman 2018).

In the 2006 plan, the term *home-grown* or *indigenous* (*zizhu chuangxin*) is used to describe a type of innovation. "Indigenous innovation," like many Chinese political slogans, is somewhat vague but as articulated in the MLP, the term refers to "genuinely original innovation; integrated innovation, the fusing together of existing technology in new ways, and 'reinnovation,' which involves the assimilation and improvement of imported technology" (Cao, Suttmeier, and Simon 2006, 40). In concrete terms, these approaches have translated to investments in the material sciences, the physical sciences, agriculture, energy, the environment, and health, as well as reforms to R&D (Cao, Suttmeier, and Simon 2006). Etymologically, *zizhu chuangxin* draws on the concept of *zili gengsheng*, a Mao-era term signifying an approach to science inspired by a nationalist drive for self-reliance that, when put into practice, essentially cut off Chinese scientists from the international research community. Cao, Suttmeier, and Simon (2006) point out that while the MLP's call for indigenous innovation does not signal a desire for total self-reliance, it is nevertheless a component of a drive to increase national autonomy and control. The impulse is central to an attempt by the state to shift China from a manufacturing to a knowledge economy and in the process develop its own intellectual property and technical standards (Lazonick, Zhou, and Sun 2016).

In terms of addressing environmental problems, the construction of large infrastructural projects such as hydropower dams (Bradsher 2017) speaks to China's conceptualizing of S&T innovation in the form of the implementation of mega-projects. During earlier periods in Chinese history, environmental issues were understood as primarily problems of overcoming scarcity and controlling the limits of nature (Shapiro 2001). Agricultural biotechnology was the state's answer to the Malthusian narrative of famine and starvation (Chen 2010). In 1988, China became the first country to grow a commercial genetically modified (GM) crop (tobacco), hinting at an enthusiasm for and investment in biotechnology (Pray 1999). However, since the Maoist period, efforts to pursue development through state-sponsored science also produced ecological disasters (Shapiro 2001) and generated widespread suspicion about the environmental impact of state-sponsored technological solutions to development. In the post-reform period, the object of environmental management is no longer material scarcity but environmental degradation, a problem exacerbated by the pursuit of economic growth over the last four decades. Air pollution, toxic waterways, and the depletion of forests and other natural resources have become key national concerns (Economy 2010) in that they threaten sustainability and health, as well as the party's ability to deliver on the promise of creating a well-off society.

In the early 2000s, the Chinese state began to respond to these concerns, and official policy shifted from a focus on economic growth measured strictly in terms of gross domestic product (GDP) toward incorporating sustainability as a metric of development, particular in Chinese cities. Since the early 2000s, words like *sustainability* or *environmental goals* have frequently shown up in policy documents. In 2002, environmental protection became a part of the official policy of building a "well-off society" (Stern 2013; Tilt 2010). Campaigns for environmental improvement have been especially important in cities, where a growing middle class increasingly demands environmental improvement. In campaigns for "blue sky days" and sanitary cities, official policy aims to remediate pollution by, for example, eliminating cars, increasing green areas, cleaning up waterways, and promoting citizen recycling. On the ground, nonstate actors such as developers and farmers have responded to official calls for sustainable development by incorporating similar ideas and terminology into their own projects (Fearnley 2015; Sze 2014).

The Problem of Waste Management in China

Unlike in the socialist era, when Chinese cities were primarily organized around production (Bray 2005), post-reform Chinese cities have become spaces for consumption (Davis 2000; Zhang 2010). By the early 2000s, a growing crisis over how to manage municipal solid waste emerged in cities across China. The amount of municipal waste produced in Guangzhou increased from 1,424 tons per day in 1984 to 11,325 tons per day in 2012.[5] This rise not only resulted from an increase in the growth of the urban population (from 4 million to 16 million) but also reflected a critical change in Chinese citizen's relationship to consumption and disposal (Goldstein 2006). While the socialist era was characterized by an ethic of frugality and thrift, in the post-reform period, consumption and its corollary, waste, speak to a wealthy and developed nation (Bogner et al. 2007) and further signal the liberation of citizens from a condition of material scarcity (Farquhar 2002).

The state's reliance on technology to solve the waste problem was instituted on the heels of campaigns to change wasteful behavior. In an effort to curb corruption, the central government put in place regulations against banquets and feasting, along with a series of austerity measures (Jacobs 2013). During the 2013 annual *chuangwen, chuangwei* (Sanitation and Civilizing City) campaign in Guangzhou, restaurants and shopping malls encouraged people to reform their own behaviors by limiting the amount of food waste they produced. Campaigns against wasteful behavior, which echo socialist austerity campaigns that targeted perceived excess, came to be expressed in the rhetoric of environmental concern. However, in spite of official edicts, public campaigns encouraging citizens to recycle remained for the most part unenforced. By the end of 2013, state recycling campaigns that pushed citizens to separate their wet from dry waste were largely a failure, and most residents simply discarded their organic waste with other refuse.

Guangzhou's preference for centralized waste treatment infrastructure reflects the state's vision of ecological modernization through technological transfer, a strategy that relies on foreign technology for waste management. From 1998 to 2005, Guangzhou's municipal waste was hauled to Asia's biggest sanitary landfill—Xinfeng landfill, in operation since 2002 and built and operated by a foreign multinational. Following in the footsteps of more developed nations such as Japan and Denmark, the city has since pursued WTE incineration as the primary mode of waste treatment. The first WTE incinerator in Guangzhou, constructed through a build-operate-transfer (BOT) scheme with the French environmental services company Veolia, began operating in 2005. WTE incineration, an imported and advanced technology, was presented as a more sustainable way to treat waste. As of 2016, in addition to the two WTE incinerator plants already up and running in the city, Guangzhou was in the process of constructing five more facilities by 2017 with a total combined capacity of 20,000 tons of waste per day (Xin 2016).

Since the proposed siting of the first WTE incinerator in the Southern District of Panyu in 2009, however, citizens have protested what they perceive to be a toxic technology that releases dioxin, heavy metals, and other pollutants into the atmosphere. Citizens mobilized against proposed WTE incinerators in Panyu, Huadu and Zengcheng, organizing online, attending petitioning meetings and protesting in the streets (Zhang 2014; Zhao 2011). As protestors built a case against WTE incineration, their investigations also revealed that a high composition of organic waste plagued the implementation of WTE technology in China. In 2013, organic waste, mostly leftover kitchen scraps, made up of over 50 percent of Guangzhou's total municipal waste stream. Against claims that WTE incinerators represented a Western and thus advanced green environmental technology, waste activists argued that the technology was ill-suited to China's waste. In 2013, the controversy over WTE incinerators led to an open search for an appropriate solution to manage organic waste (Qiu 2016).[6]

As of 2013, Guangzhou lacked a viable technology to sustainably process urban organic waste. In the premodern, socialist, and even early post-reform era cities, animals constituted an informal organic waste treatment system, as chickens or ducks roamed in neighborhoods and gardens, and local hogs feasted on kitchen scraps and leftovers. Today, most urban residents live in planned high-rises, and chickens, ducks, hogs, and rabbits are viewed as markers of a backward and rural environment, to be evicted from China's modern urban centers. In a quest to realize a vision of modernity characterized by a clear distinction between the urban and the rural, working animals were cast out of cities and with them a system of organic waste management.

At that time, composting and bio-gas facilities, both existing technologies under consideration by officials in Guangzhou, had yet to prove feasible for the large-scale treatment of urban organic waste. The Datianshan composting pilot project, which processes over 100 tons of organic waste a day, was quickly losing government support. Originally conceived of as an organic waste treatment facility, a garden, and an educational facility for farming, the project was widely popular with the public. However, by 2013, it had fallen out of favor with the lead engineers of waste management in Guangzhou, who critiqued it on economic grounds and questioned the marketability of soil made from composted organic waste (Huang 2013).

Bio-gas treatment, otherwise known as *anaerobic digestion*, was another technology commonly used for organic waste. Anaerobic digestion uses microorganisms to break down organic compounds in the absence of oxygen. Compared with composting, which releases carbon dioxide with little value, anaerobic digestion produces energy. In the 1960s and 1970s, the government subsidized the installation of both household digesters and centralized plants across the Chinese countryside.[7] Gas produced from the digester is used primarily for cooking and to supplement electricity and to run generators. The digested slurry and sludge also serve as an effective form of fertilizer to cultivate earthworms, as feed for poultry and fish, or to grow mushrooms and lotus roots (Li 1984). While small-scale bio-gas facilities were popular with rural households, the idea of using bio-gas for organic waste in Guangzhou in 2013 was just getting off the ground. Commonly used to treat and reduce the volume of wastewater and sewage sludge, bio-gas technology has seen only limited use as a treatment for other forms of organic municipal waste. Municipal government proposals to introduce bio-gas digesters typically involved building high-tech facilities next to existing waste management facilities such as WTE incinerators.

During my time in the field, the problem of organic waste motivated scientists, entrepreneurs, and environmentalists to experiment with a wide range of small-scale, local solutions for organic waste management, rather than looking exclusively to foreign technology. I visited companies trying to introduce garbage disposal units into Chinese households and developing local shredders and processors, as well as environmentalists brewing and fermenting organic household cleaning products (Zhang 2018). Dr. Wu believed the BSF to be an ideal, environmentally sound technology to confront the technical problem of organic waste, which also promised to generate economic value.

During my fieldwork, I watched as Dr. Wu experimented with a pilot project to breed BSF in an experimental laboratory. Two years later, he received state funding to develop different prototypes for the BSF project. By 2015, in addition to Dr. Wu's pilot project, districts in Huadu and Baiyun were in the process of setting up independent programs to use the BSF to treat between 200 and 400 tons of organic waste per day, a greater capacity than any previous organic treatment scheme (Liang and Cheng 2015). By 2018, a local NGO had begun a collaboration with a private sanitation company to use the BSF to treat locally collected organic waste. Six years after I first encountered Dr. Wu's pilot project, a niche environmental research project was beginning to achieve recognition as a viable technology for organic waste.

The Black Soldier Fly Project as Chinese Science

The BSF project represents a turn toward indigenous innovation in China not only insofar as it emerged in response to local challenges, but also because the technology is reinforced by a Chinese tradition of entomological research that understood insects as potentially useful. Dr. Wu's early research into the BSF was based at the Guangdong Entomological Institute (GEI), originally established in 1958 as a part of the Chinese Academy of Sciences. Entomological research was tightly linked to agricultural development in China and the GEI, as well as the South China Agricultural University and the Guangdong Academy of Agricultural Sciences, are prominent examples of institutions with a long and rich tradition of entomological research in Guangdong. In 2013, GEI researchers were actively engaged in research projects concerned with the protection and use of wild animals, the development of biotechnology, and both agrarian and urban pest management (Guangdong Entomological Institute, n.d.).

As Sigrid Schmalzer shows, Chinese and Western entomology diverged during the socialist period as Chinese researchers focused on the development of biological pest control. Since American entomology was particularly influenced by chemical corporations and moved toward a reliance on chemical pesticides, U.S. delegates visiting the Big Sand Commune in Guangzhou during the mid-1970s marveled at the experiments they saw in biological pest control, which featured "bug-eat-bug" and microbial approaches (Schmalzer 2016). The pilot project at Big Sand, championed by U.S.-educated entomologist Pu Zhelong, was part of a turn toward devising indigenous and ecological scientific solutions, in line with a more nativist, proletarian approach to scientific research that Schmalzer characterizes as uniquely Chinese, one that emphasizes "the 'earthy values' of a *tu* science" (Schmalzer 2016, 59).

China has a long history of cultivating, using, and living with insects. Silkworms in particular have a long history in China's economic production. The mulberry plot–fishpond, a system of Chinese silk production practiced in the Pearl River Delta and Yangtze River Delta since 3500 BC, used silk worms to sustain a lucrative textile industry (Min and Hu 2001). On silk farms, carp was raised on silkworm feces in ponds, the sludge from which fertilized the mulberry plants that silkworms relied on. Insects have also long been an integral part of China's urban life. For example, the cultivation of crickets, which were companions in gambling houses, was an ancient art form (Raffles 2010). Illustrative of what Lu Xun calls a "primitive acceptance of insects," AhQ—the protagonist in the short story *The True Story of AhQ* (1921)—exhibits a causal acceptance of a lice infection on his body, a reminder of the once intimate relationship between humans and insects.

But as Ruth Rogaski observes, beginning in the modernist period, living among insects, mosquitos, flies, and lice was perceived by the state as an indication of the deficiencies of the social body. Modernist reformers in Republican China saw this particular relationship with insect as an index of other politically backward characteristics of being Chinese, such as being "illogical" and "belligerent" (2002, 398). The establishment of public health discourse instilled an understanding of insects as vectors of disease that threatened the health of the social body.

In socialist China, experiments with ecologically focused entomology coincided with a growing push to eradicate insects from urban centers as a part of sanitation campaigns. Public health campaign in the 1950s enacted an explicit political project of reshaping human-insect relations, which was intended to lift China into a form of political and social modernity, particularly in urban centers. The birth of China's first public health campaigns can be traced back to fear of biological warfare. The Communist state suspected that diseased rats were dropped from planes by the United States in Gannan village in Heilongjiang Province near North Korea (Rogaski 2002). The complete eradication of flies, rats, and even dogs in the village became a model for China's first public health campaign, the Patriotic Hygiene Campaign (*Aiguo weisheng yundong*), which advocated the killing of all pests, along with the dousing of every villager with a 5 percent solution of DDT. This campaign, along with other subsequent health campaigns, established a theme for the relationship between insects, public health, and modernity that endures in Chinese society; Rogaski characterizes it as "The need to eradicate, to exterminate, or to annihilate (*chu, xiaomie*) perceived enemies, whether political or natural, in order to achieve a state of modernity" (2002, 382). The elimination of "pests" on a mass scale in order to increase China's efficiency, sanitation, and modernity was again on display in the more extreme 1958 campaign to "Eliminate the Four Pests"—rats, sparrows, flies, and mosquitoes—in order to "eradicate disease, lift the spirits, change and improve habits, and transform the nation" (Bao 2012, 321–322). Quotas and rewards were set by local officials for the killing of swallows and flies. Later, when scientists presented Mao with evidence of the benefits of swallows, the eradication campaigns continued with bedbugs replacing swallows on the list (Bao 2012, 322).

The perception of common urban insects such as flies and mosquitos as pests that needed to be eradicated continued into the late Mao period and pervaded contemporary urban health discourse. In 1974, when A. J. Smith published his observations on public health in China in the *British Medical Journal*, he noted that one of the astonishing results of the Elimination of the Four Pests Campaign was that "the housefly is virtually extinct in China" (Smith 1974, 492). In Guangzhou's subtropical climate, public health campaigns against dengue fever and malaria have also commonly targeted mosquitoes. Since 1978, public health campaigns have increasingly moved toward a series of competitions and rewards between cities under the "Hygienic City Campaign." The competition has five basic criteria: garbage disposal management (i.e., urban garbage disposal rate of no less than 80 percent), sewage treatment, greening, air quality and finally, pest control (Zhang and Li 2011). The sanitary discourse pits the health and survival of insect and human population against one another in a way that makes the need to segregate and eliminate insects from human environments seem not only natural but necessary.

In this context, the capacity to imagine insects not only as a vector of disease but also as a tool for waste management echoes an earlier tradition in Chinese entomological research to treat insects not just as a population to be controlled but simultaneously as a form of technology and resource for other purposes. For Dr. Wu, his work with the BSF dates back to his doctoral dissertation focused on potential uses for the BSF. Along with fellow researchers, he has secured a number of patents related to the standardization of fly-raising into a replicable and value-generating technology. For instance, his patent includes a piece of specialized equipment designed to separate BSF pupae, an apparatus to house BSF larvae and a technique for extracting *chitosan*—a chemical compound that is usually obtained from crustacean shells that can be used in agriculture and biopesticides—from the BSF.[8]

My first visit to the Entomological Institute began with a formal presentation on the BSF by Director Luo,[9] the head of the institution:

> I'm sure that you've all learned about the BSF from the web, but let me do a brief introduction . . . From the beginning, we've designated the BSF a technology, one that is useful for solid organic waste management. Compared with the conventional composting technology, there are many advantages. BSF uses an insect for processing. This has a higher efficiency than microorganisms. Why? Because the BSF feeds on solid organic waste. The process of BSF feeding on organic waste is not one of decomposition (fenjie), but actually a process of transformation (zhuanhua), one of transforming [organic waste] into animal protein. Insect protein is a high value-added resource (gao fujiaze zhiyuan). Microorganisms break down and decompose elements into a more simplified [compound], useful only for composting. By using the BSF to treat kitchen waste, once it matures, we can process it into all kinds of useful products (jiagong chen gezhong ge yang de you yi de). [Fieldnotes 2012]

The meeting took place at a conference room in the research institute. The PowerPoint presentation was filled with references to academic studies, data and statistics, providing a scientific framing to the project. In his presentation, Luo foregrounds the biological life-cycle of the BSF explicitly as a technology undergoing active research and development. Under this rubric, the digestive system of the BSF is presented as a mechanized process with a higher efficiency for transforming an underutilized product into a "value-added resource." The rhetoric of "organic waste recovery" translates the cycling of nutrients between the rural and the urban sphere according to a scientific and economistic register to imply maximum efficiency. The body of the BSF is relegated to an animal protein and a commercial product that can be capitalized on through various value-added processes.

In addition to experimenting with methods of breeding at the GEI, Dr. Wu also focused on creating a new protein from insects. When I returned to the field in 2018, Dr. Wu's laboratory was rearranged into a temporary workshop to turn the BSF into a product. While he continued to experiment with designing technologies to rear flies for waste management purposes, the main focus of his attention has shifted to the processing and marketing of fly larvae as a viable commercial feed. While his earlier experiments were focused on basic research and funded by institutional grants, including a grant from the prestigious Chinese Academy of Science, since 2013, his work has taken a more applied and practical turn as he tried to raise capital in order to scale up production, and uncover new markets for his product. During a conversation in late 2013, Dr. Wu explained to me that opportunities for scientific innovation in China are limited compared to the enviable condition for technological development in other developed nations, best exemplified by Silicon Valley (Field notes 2013). Development in the West, he informed me, is guided by idealism while in China there is "only pragmatism" (Field notes 2013). He admired most about the West what he described as a spirit of entrepreneurialism in which scientific discoveries can blend seamlessly with commercial value.

As Greenhalgh argues in the introduction to this volume, efforts to transform China into an innovative nation have broadened beyond scientific research and development and that scholars should pay close attention to the relationship between scientific research and market forces, entrepreneurship and financing structures. Wu is keenly attuned to this tension between primary research on the one hand and the need, on the other, to devise technical solutions that can be easily and effectively marketed and commodified. Returning to his laboratory in 2018, I watched as his assistants weighed and packaged bags of protein to be sold to a shrimp farm. Each morning, Wu purchases fly larvae from a local laboratory and works on pulverizing the fly larvae to combine them with additives. Dr. Wu framed the product as a "nutritional supplement" for shrimp. Over lunch he calculated how quickly the speed of production must increase to match the rate of demand and strategized on how to raise the necessary capital to scale up production. He explained to me that these days, he tends to think of himself not only as a scientist but also as an entrepreneur whose role is oriented less toward scientific discovery and more toward the development of a marketable product (Field notes 2018).

Innovating the BSF as a solution for organic waste forced Wu to leave behind the world of basic research and scientific experimentation to enter the world of applied science, where entrepreneurial scientists are centrally concerned with the adaption of research to market forces. Scholars argue that "innovating through commercialization"—bring products quickly to market then iterate based on consumer wants—embodies a particularly Chinese approach to innovation (Economy 2018). In 2018 another group of scientists at a nearby research university also working with the BSF formed a company that both sells BSF larvae as animal feed to Europe and the U.S., and to export the system of BSF raising to other developing markets. At the time that I visited in 2018, the company proudly shared with me the successful franchise of their system to a farm in Nigeria. China's development as an innovation nation and the new focus on "indigenous innovation" is inseparable from a desire to pose local scientific and technological development as not only forms of knowledge that can be circulated but also to create a knowledge product that can quickly be developed for export markets. Indigenous innovation parallels the attempt to shift China from the recipient to the donor of technological innovation.

Living with Insects

In China, a shift in official policy alongside a need for technical solutions to address local problems have generated renewed investment in scientific and technological innovations. The implications of S&T, however, are not limited to the realm of policy-making or engineering. The ramifications of using the BSF as a waste management technology are ecological as well as social and have consequences for humans as well as non-humans. As a waste management technology, the BSF is no longer an insect in the wild but a socio-technical hybrid devised to carry out a human purpose. Recast as an environmental technology and a technical innovation, the BSF generates new ways that humans use, live with and become increasingly entangled with the survival of another species.

Environmental anthropologists emphasize the important role that animals play in mediating social, economic and environmental relations (Evans-Pritchard 1969; Rappaport 2000). Recent research illustrates the role of animals not only as vessels of cultural meaning but also emphasize the social relations generated between humans and animals at particular sites and between particular species (Ogden, Hall, and Tanita 2013, 6). Ethnographers point to "contact zones" where "encounters between *Homo Sapiens* and other beings generate mutual ecologies and coproduced niches" (Kirksey and Helmreich 2010, 546). Multispecies ethnographies emphasize moments of "reciprocal capture" where humans and nonhumans establish a more equitable role and where humans hold an interest in seeing the others maintain its existence (Kirksey 2013). Studies of mammals and other large, charismatic species such as whales (Neves 2010) and primates (Fuentes and Wolfe 2002) stress their ability to elicit ethical obligations and feelings of mutual vulnerability between humans and animals, especially during moments of interaction and recognition (Parreñas 2012). However, other species are bred to exclusively serve human needs. Animals used for agricultural production have been cast in relations of domestication, experimentation, and subject to engineering as new forms of technologies and labor (Blanchette 2015; Parikka 2010). With the exception of scholars writing on conservation (Coggins 2003; Hathaway 2013) or on the ways that animals, particularly insects, have been cast as vectors of infectious disease and as central players in viral outbreaks and epidemics (Fearnley 2015; Zhan 2005) there is a dearth of ethnographies of non-human life in China.

Sanitation campaigns to eradicate urban pests show how the creation of modern urban environments dramatically altered ecological relations between humans and other forms of life. Hygienic modernity, dominated by concerns about sanitation, resulted in both the ejection from and systematic control of animals in cities (Rogaski 2004). In tropical climates, insects are particularly targeted as vectors of disease. Mosquitos have been an integral focus of tropical medicine aimed at rooting out *dengue*, yellow fever and malaria, either through the elimination of sites of mosquito larvae or the chemically intensive approach of spraying adulticide to control adult populations (Shaw, Robbins, and Jones 2010). Similarly, in a study of campaigns to rout out urban mosquitos in Dar Es Salaam, Kelly and Lazaun argue that public health interventions can be conceptualized as a "labor of disentanglement" to separate insects that have long thrived in human environments. They point out that modern ideas of urban "civic-ness" entail a relationship of distance between mosquitos and humans (Kelly and Lezaun 2014). Eradication campaigns become extensions of how people exercise their "imagination of community building" to determine which species and humans can be included in the political community (Kelly and Lezaun 2014)

The success of the BSF project hinges on the introduction of an insect population back into a contemporary, modern city. The result is that an environmental imperative couple the sustainability of an urban population with the reproduction of non-human, insect populations. As a biotechnology, BSF renders a surplus matter (organic waste) into a surplus population (BSF insects) and makes the life-cycle of the BSF dependent on cyclical output of waste produced by humans. According to Aihwa Ong, the convergence of neoliberalism and science configures "the risk of 'surplus population' and 'surplus needs' into biotech opportunities for growth" (Ong 2010, 5). The BSF project reflects a desire for scientific projects to generate market opportunities to ensure economic growth. Surplus organic matter produced by humans have become integrated with the reproductive life-cycle of insects, generating a surplus of biological organisms that can then be digested and marketed as economic profit. However, Dr. Wu's attempts to breed the BSF for urban waste management suggests that the pursuit of technological innovation for sustainability has the capacity to generate new ecological dependencies between humans and other species. Rather than a relationship of inter-species disentanglement, the project of building a sustainable modernity in China is also a nascent project that is formed through "becoming-with" (Haraway 2016) insects, in which human ecological well-being has come to depend on a closer integration with insects, but in which non-human life is increasingly captured for the generation of value. The BSF shows that dreams of national rejuvenation have created new roles and interventions for non-human populations. In an era of environmental vulnerability, S&T play a role in efforts to reproduce, improve and rationally control non-human populations in the service of ecological goals, but in doing so, such projects make human sustainability increasingly dependent on other life-forms.

Conclusion

Since the late-Qing period, technological and scientific innovation was broadly understood as the adoption of western knowledge and technology. However, the Xi government now positions China as an emerging "innovation nation" with a capacity for indigenous innovation. In addition to top-down programs to generate innovation, local scientists are also devising scientific and technological solution to address program across a wide range of sectors and scales outside stated policy and programs. The BSF project reveals how the pursuit of environmental sustainability pushed scientists away from a reliance on innovation through technological transfer toward projects and solutions that both emerge from and aim to address local conditions. Responding to the difficulty of treating a large composition of organic waste in the waste stream, a local problem that WTE incineration is ill-equipped to handle, entomologists devised a strategy for the sustainable management of waste that echoes an indigenous (*tu*), native and ecological approach to science and research and that also perpetuates a unique and longstanding interest in insects in Chinese culture.

The BSF project echoes a longer history in which insects were cultivated for human use. Returning to his field laboratory in 2018, I watched as Dr. Wu began to devise the BSF larvae as a marketable product. His identity as an "entrepreneur" illustrates the degree in which the push for indigenous innovation is situated at the intersection of science and the market. The increasing emphasis on entrepreneurial science speaks to the ways that a push toward "innovation" represents one way in which scientific research is also connected to the potential marketability of technical products.

This unusual waste management technology, however, not only exemplifies the changing direction of S&T research in China but also carries implications for how to understand the impact of new scientific innovations on the formation of the urban environment. In China's quest for a form of green modernity, the BSF project has the potential to generate a new relationship between human and insects. As a scientific innovation to address the problem of the sustainable management of waste, the BSF represents a population and species that humans are beginning to invest in and might come to depend on for not only waste management but ecological well-being. The cultivation of insects complicates the expectations of a sanitary approach to cities, where waste and insects are both viewed as vectors of disease and must be eradicated from the urban sphere. Instead, the project of raising and using flies, begs the question of how the desire to realize a vision of sustainable development through S&T can also generate new forms of interspecies dependence.

NOTES

1. Wu is a pseudonym.

2. The high protein animal larvae has become especially marketable since the price of fish feed increased dramatically in the last ten years both due to a boom in aquaculture and a depleted wild fish stock ("Black Soldier Fly (*Hermetia Illucens*)," n.d.).

3. Only a handful of companies in the world, such as Enterra in Canada, Enviroflight in the United States, and JM Green in China, has facilities to process more than sixty tons of organic waste per day; other experiments remain as limited backyard projects.

4. Shanzhai production was made possible only through availability and access to a wide range of manufacturing and design facilities that generated the conditions to quickly realize and revise different iterations of product design.

5. The data comes from the Guangzhou Statistical Yearbook (1984–2013). The total municipal waste was 520,000 tons/year in 1984 compared with 413,350,000 tons/year in 2012. These figures however, do not account for changes to the municipal boundaries in Guangzhou. Since 2001, figures account for Guangzhou's ten districts. The official figure commonly cited by the Guangzhou municipal management bureau for 2013 is 18,000 tons/day (Tan 2014). See http://webcache.googleusercontent.com/search?q=cache:4v5319sNV6YJ :chinagev.org/index.php/meirijianhe/4234-qiyueshisiridaodu&hl=en&gl=us&strip =1&vwsrc=0.

6. In 2016, Guangzhou municipal government announced a centralized plan for the treatment of organic waste (see Qiu 2016).

7. Over 7 million digesters were built in the countryside, used primarily to treat animal manure and other agricultural byproducts such as grain husks and stocks (see Li 1984).

8. Guangdong Entomological Institute, n.d.

9. Luo is a pseudonym.

References

Andreas, Joel. 2009. *Rise of the Red Engineers: The Cultural Revolution and the Origins of China.* Stanford, CA: Stanford University Press.

Bao, Maohong. 2012. "Environmentalism and Environmental Movements in China since 1949." In *A Companion to Global Environmental History*, edited by John Robert McNeill and Erin Stewart Mauldin, 474–492. Chichester, UK: Wiley.

Berman, Robby. 2018. "China Takes the Lead in Science and Engineering Research Publications | Big Think." Bigthink.com. http://bigthink.com/robby-berman /china-tops-the-us-in-science-publications-for-the-first-time.

"Black Soldier Fly (*Hermetia Illucens*)." N.d. Organic Waste Solutions. http://www .organicvaluerecovery.com/soldier_fly/soldier_fly.htm (accessed March 16, 2016).

Blanchette, Alex. 2015. "Herding Species: Biosecurity, Posthuman Labor, and the American Industrial Pig." *Cultural Anthropology* 30(4): 640–669. https://doi.org /https://doi.org/10.14506/ca30.4.09.

Bogner, Jean (coordinating lead author), Mohammed Abdelrafie Ahmed Sudan, Cristobal Diaz, M. Abdelrafie Ahmed, Cristobal Diaz, Andre Faaij, Qingxian Gao, et al. 2007. "Waste Management in Climate Change 2007: Mitigation." Contribution of Working Group III to the Fourth Assessment Report of the Intergovernmental Panel on Climate Change. Cambridge, UK: Cambridge University Press. https://www.ipcc.ch/pdf/assessment-report/ar4/wg3/ar4-wg3-chapter10.pdf.

Bradsher, Keith. 2017. "China Looks to Capitalize on Clean Energy as U.S. Retreats." *New York Times*, June 5. https://www.nytimes.com/2017/06/05/business/energy -environment/china-clean-energy-coal-pollution.html.

Bray, David. 2005. *Social Space and Governance in Urban China: The Danwei System from Origins to Reform.* Stanford, CA: Stanford University Press.

Cao, Cong, Richard P. Suttmeier, and Denis Fred Simon. 2006. "China's 15-Year Science and Technology Plan." *Physics Today* 59(12): 38–43. https://doi.org/10 .1063/1.2435680.

Chen, Nancy N. 2010. "Feeding the Nation: Chinese Biotechnology and Genetically Modified Foods." In *Asian Biotech: Ethics and Communities of Fate*, edited by Aihwa Ong and Nancy N. Chen, 81–94. Durham, NC: Duke University Press.

Coggins, Chris. 2003. *The Tiger and the Pangolin: Nature, Culture, and Conservation in China.* Honolulu: University of Hawai'i Press.

Davis, Deborah. 2000. *The Consumer Revolution in Urban China.* Berkeley: University of California Press.

Economy, Elizabeth C. 2010. *The River Runs Black: The Environmental Challenge to China's Future.* Ithaca, NY: Cornell University Press.

Economy, Elizabeth C. 2018. *The Third Revolution: Xi Jinping and the New Chinese State.* New York: Oxford University Press.

Elman, Benjamin A. 2006. *A Cultural History of Modern Science in China*. Cambridge, Mass: Harvard University Press.

Evans-Pritchard, E .E. 1969. *The Nuer: A Description of the Modes of Livelihood and Political Institutions of a Nilotic People*. New York: Oxford University Press.

Fan, Fa-ti. 2012. "'Collective Monitoring, Collective Defense': Science, Earthquakes, and Politics in Communist China." *Science in Context* 25 (01): 127–154. https://doi .org/10.1017/S0269889711000329.

Farquhar, Judith. 2002. *Appetites: Food and Sex in Postsocialist China*. Durham, NC: Duke University Press.

Fearnley, Lyle. 2015. "Wild Goose Chase: The Displacement of Influenza Research in the Fields of Poyang Lake, China." *Cultural Anthropology* 30(1): 12–35. https://doi .org/https://doi.org/10.14506/ca30.1.03.

Fuentes, Agustín, and Linda D. Wolfe. 2002. *Primates Face to Face: The Conservation Implications of Human-Nonhuman Primate Interconnections*. Cambridge, UK: Cambridge University Press.

Goldstein, Joshua. 2006. "The Remains of the Everyday: One Hundred Years of Recycling in Beijing." In *Everyday Modernity in China*, edited by M. Y. Dong and Joshua Goldstein, 260–302. Seattle: University of Washington Press.

Guangdong Entomological Institute. N.d. "About Us." http://www.gkbee.com/jianjie/.

Hathaway, Michael J. 2013. *Environmental Winds: Making the Global in Southwest China*. Berkeley: University of California Press.

Haraway, Donna J. 2016. *Staying with the Trouble*. Durham, NC: Duke University Press.

Huang, Shaohong. 2013. "Datianshan xunhuan shengtaiyuan canchu laji xiangmu tingchan zuo jieduanxing tiaozheng (Datianshan Project Halted for Adjustments)." *Southern Daily*, October 28. http://gz.fzg360.com/news/201310/466265_1.html.

Jacobs, Andrew. 2013. "Elite in China Face Austerity under Xi's Rule." *New York Times*, March 27. https://www.nytimes.com/2013/03/28/world/asia/xi-jinping-imposes -austerity-measures-on-chinas-elite.html.

Kelly, Ann H., and Javier Lezaun. 2014. "Urban Mosquitoes, Situational Publics, and the Pursuit of Interspecies Separation in Dar Es Salaam." *American Ethnologist* 41(2): 368–383. https://doi.org/https://doi.org/10.1111/amet.12081.

Kirby, William. 2000. "Engineering China: Birth of the Developmental State, 1928– 1937." In *Becoming Chinese: Passages to Modernity and Beyond*, edited by Wen-Hsin Yeh, 137–160. Berkeley: University of California Press.

Kirksey, S. Eben. 2013. "Interspecies Love: Being and Becoming with a Common Ant, Ectatomma Ruidum (Roger)." In *The Politics of Species: Reshaping Our Relation- ships with Other Animals*, 164–176. Cambridge, UK: Cambridge University Press.

Kirksey, S. Eben, and Stefan Helmreich. 2010. "The Emergence of Multispecies Ethnog- raphy." *Cultural Anthropology* 25(4): 545–576.

Lazonick, William, Yu Zhou, and Yifei Sun. 2016. "Introduction: China's Transforma- tion to an Innovation Nation." In *China as an Innovation Nation*, edited by Yu Zhou, William Lazonick, and Yifei Sun, 1–32. New York: Oxford University Press.

Li, Nianguo. 1984. "Biogas in China." *Trends in Biotechnology* 2(3): 77–79.

Liang, Yitao, and Guangwei Cheng. 2015. "Guangzhou Huadu the First District to Experiment with Using the Black Soldier Fly to Treat 200/Ton of Organic Waste (Yong heishuimeng chi canchu laji Guangzhou Huadu xianxingxianshi meitianchi 200 dun laji)." *Jinyang Web*, May 21. http://news.ycwb.com/2015-05/21/content _20226231.htm/.

Lindtner, Silvia. 2014. "Hackerspaces and the Internet of Things in China: How Makers Are Reinventing Industrial Production, Innovation, and the Self." *China Information* 28(2): 145–167. https://doi.org/10.1177/0920203X14529881.

Lindtner, Silvia, Anna Greenspan, and David Li. 2015. "Designed in Shenzhen: Shanzhai Manufacturing and Maker Entrepreneurs." *Aarhus Series on Human Centered Computing* 1(1): 12. https://doi.org/10.7146/aahcc.v1i1.21265.

Macur, Juliet. 2007. "Beijing Air Raises Questions for Olympics." *New York Times*, August 26. http://www.nytimes.com/2007/08/26/sports/othersports/26runners.html.

Min, Kuanhong, and Baotong Hu. 2001. "Chinese Embankment Fish Culture." In *Integrated Agriculture-Aquaculture: A Primer*, edited by ICLARM—the World Fish Center, International Institute of Rural Reconstruction, and Food and Agriculture Organization of the United Nations. Rome: Food and Agriculture Organization. http://www.fao.org/docrep/005/y1187e/y1187e00.htm#TopOfPage.

Neves, Katja. 2010. "Cashing in on Cetourism: A Critical Ecological Engagement with Dominant E-NGO Discourses on Whaling, Cetacean Conservation, and Whale Watching." *Antipode* 42(3): 719–741.

Ogden, Laura A., Billy Hall, and Kimiko Tanita. 2013. "Animals, Plants, People, and Things: A Review of Multispecies Ethnography." *Environment and Society: Advances in Research* 4(1): 5–24.

Ong, Aihwa. 2010. "Introduction: An Analytics of Biotechnology and Ethics at Multiple Scales." In *Asian Biotech: Ethics and Communities of Fate*, edited by Aihwa Ong and Nancy N. Chen, 1–54. Durham, NC: Duke University Press.

Parikka, Jussi. 2010. *Insect Media: An Archaeology of Animals and Technology*. Minneapolis: University of Minnesota Press.

Parreñas, Rheana "Juno" Salazar. 2012. "Producing Affect: Transnational Volunteerism in a Malaysian Orangutan Rehabilitation Center." *American Ethnologist* 39(4): 673–687. https://doi.org/https://doi.org/10.1111/j.1548-1425.2012.01387.x.

Pray, Carl E. 1999. "Public and Private Collaboration on Plant Biotechnology in China." *AgBioForum* 2(1): 48–53.

Qiu, Ping. 2016. "Guangzhou Municipal Management Bureau: Plans to Build and Expand Facilities to Treat 4,800 Tons of Organic Waste/Day (Guangzhou chengguanwei: Shejiyiri Chuli 4800 dun canchulaji, haixiangkuoda)." *Southern Metropolitan Daily*, January 16. http://news.oeeee.com/html/201601/16/359553.html.

Raffles, Hugh. 2010. *Insectopedia*. New York: Pantheon.

Rappaport, Roy A. 2000. *Pigs for the Ancestors: Ritual in the Ecology of a New Guinea People*. Prospect Heights, IL: Waveland Press.

Rogaski, Ruth. 2002. "Nature, Annihilation, and Modernity: China's Korean War Germ-Warfare Experience Reconsidered." *Journal of Asian Studies* 61(2): 381–415.

Rogaski, Ruth. 2004. *Hygienic Modernity: Meaning of Health and Disease in Treaty-Port China*. Berkeley: University of California Press.

Schmalzer, Sigrid. 2016. *Red Revolution, Green Revolution: Scientific Farming in Socialist China*. Chicago: University of Chicago Press.

Shapiro, Judith. 2001. *Mao's War against Nature: Politics and the Environment in Revolutionary China*. Leiden: Cambridge University Press.

Shaw, Ian Graham Ronald, Paul F. Robbins, and John Paul Jones. 2010. "A Bug's Life and the Spatial Ontologies of Mosquito Management." *Annals of the Association of American Geographers* 100(2): 373–392.

Shen, Grace Yen. 2014. *Unearthing the Nation: Modern Geology and Nationalism in Republican China*. Chicago: University of Chicago Press.

Smith, A. J. 1974. "Medicine in China: Public Health in China." *British Medical Journal* 1 (June): 492–494.

Stern, Rachel E. 2013. *Environmental Litigation in China: A Study in Political Ambivalence*. Cambridge, UK: Cambridge University Press.

Sze, Julie. 2014. *Fantasy Islands: Chinese Dreams and Ecological Fears in an Age of Climate Crisis*. Oakland: University of California Press.

Tan, Zheng. 2014. Guangzhou qidong chuangjian quanguo shenghuo laji fenlei shifan chengshi (Guangzhou creates national recycling pilot project). CRI Online. July 10, 2014. http://webcache.googleusercontent.com/search?q=cache:4v5319sNV6YJ :chinagev.org/index.php/meirijianhe/4234-qiyueshisiridaodu&hl=en&gl=us&strip =1&vwsrc=0.

Tilt, Bryan. 2010. *The Struggle for Sustainability in Rural China: Environmental Values and Civil Society*. New York: Columbia University Press.

Wei, Chuanjuan Nancy, and Darryl E. Brock, eds. 2013. *Mr. Science and Chairman Mao's Cultural Revolution: Science and Technology in Modern China*. Lanham, MD: Lexington Books.

Wei, Ren. 2016. "Beijing Aims to Lead World in Innovation by 2050." *South China Morning Post*, May 20. http://www.scmp.com/news/china/policies-politics/article /1947968/beijing-aims-lead-world-innovation-2050.

Xin, Junqing. 2016. "'Five More WTE Incinerator in the Midst of Construction and an Additional Planned in Guangzhou.'" (Guangzhou laji fenshao fadianchang 5 zuo zai jian 1 zuo lixiang)." *Southern Daily*, Sept. 12. http://www.cnenergy.org/hb /201609/t20160912_376694.html.

Zhan, Mei. 2005. "Civet Cats, Fried Grasshoppers, and David Beckham's Pajamas: Unruly Bodies after SARS." *American Anthropologist* 107(1): 31–42.

Zhang, Amy. 2014. "Rational Resistance: Homeowner Contention Against Waste Incineration in Guangzhou." *China Perspectives* 2: 46–52.

Zhang, Amy. 2018. "(Eco)Enzyme as Catalyst." Theorizing the Contemporary, *Fieldsights*, July 26. https://culanth.org/fieldsights/ecoenzyme-as-catalyst.

Zhang, Chris. 2017. "Can Red China Really Be the World's New Green Leader?" *Diplomat*, November.

Zhang, Li. 2010. *In Search of Paradise: Middle-Class Living in a Chinese Metropolis*. Ithaca, NY: Cornell University Press.

Zhang, Yongmei, and Bingqin Li. 2011. "Motivating Service Improvement with Awards and Competitions—Hygienic City Campaigns in China." *Environment and Urbanization* 23(1): 41–56.

Zhao, Katherine. 2011. "Boundary-Spanning Contention: The Panyu Anti-Pollution Protest in Guangdong, China." *Stanford Journal of East Asian Affairs* 11(1):17–25.

Zhu, Sheng, and Yongjiang Shi. 2010. "*Shanzhai* Manufacturing—an Alternative Innovation Phenomenon in China." *Journal of Science and Technology Policy in China* 1 (1). Emerald Group Publishing Limited:29–49. https://doi.org/10.1108 /17585521011032531.

UNMASKING A GENDERED MATERIALISM

Air Filtration, Cigarettes, and Domestic Discord in Urban China

Matthew Kohrman

> **When it's so grey outside, my husband tells me I should head home and hunker down next to our air purifier.**
>
> —Chen Xitao

> **They dance those lights. They twinkle. My husband turns them on whenever it's bad outside. Does he believe in them? I don't know. What I know . . . is that they make me feel angry and trapped.**
>
> —Ma Li

> **PM2.5 goes up, my mood drops. The purifiers help. I always keep them running at home. But I can't stay home all the time. That would make me feel even worse. And of course, I have to go outside to smoke. Annoying. But I have learned not to fight with my wife over that. No cigarette smoking in the house. Right? But if we have the air purifier running, what's the big deal?**
>
> —Zhou Jun

Fraught feelings about fibrous filters are on the rise in China. As they experience environmental contamination, people are reaching for old and new ways to envelop themselves in the materiel of cleanliness, purity, and unadulteration. A teleology of purity is hardly new to China (Rogaski 2004), but how residents of the People's Republic are now striving to create microclimates of cleanliness for themselves and their intimates seems to be changing dramatically—and so is the emotional toll of making and sustaining slipstreams of unadulteration. Newly named specters from *airpocolypse* to *airmageddon* and *cancer villages* (Lora-Wainwright 2010) prompt cycles of affect and action. Urgencies and practices converge around how and whether to filter out contaminants from what one breaths, swallows, and touches. To profit-seekers, this convergence is an unrivaled opportunity. If the most dependable money-makers during the American gold rush were shovels, sieves, and scales, today in China the shrewd opportunists are those flogging filtration—the filter itself, what it delivers, and all manner of instrument to measure that delivery. Air purifiers, face masks, kitchen exhausts, low-VOC paint, bottled water, specrometers, detectors, microscopes, pollution monitoring apps, and organic foods are just a few examples. These are just some of the latest technologies to animate the bromide of science saving China.

But what else should we make of all the air filtration sweeping China? Is it simply another example of risk society in action (Beck 1992, 1999; Kohrman 2004)? And within that, is it simply one more Chinese Communist Party–sanctioned neoliberalish response to a large governmentally generated mess? Sounds plausible. To be sure, personalized filtration technologies hold the promise for the Communist regime of defusing a potentially destabilizing problem by encouraging people to bury their heads in rhetorics of personal responsibility. Rather than challenging the political system, buying and deploying filters to screen out daily pollutants is what Beijing authorities would prefer heath conscious citizens do (Kay, Zhao, and Sui 2015).

Memes and Masks

Although that line of analysis has many merits, there are other matters that I want to consider here in this preliminary discussion of air filtration in contemporary Chinese contexts. And to that end, allow me to introduce an ongoing thread of internet communication. It draws on a genre—comedy—that the party has at times indicated it would prefer filtered from circulation (Magnier and Lin 2016). A stock-in-trade of humorists everywhere is using material culture to skewer common public concerns. So it should be of little surprise to see that China's bubbling internet has been generating humorous memes that invoke popular anxieties sitting at the crossroads where current Premier Li Keqiang announces a "War on Pollution" and President Xi Jinping touts his aspirational "China Dream." An iconic prong of this humorous handiwork are memes that comingle two common objects: the surgical-style respiratory mask and the cigarette. While permutations of these memes have circulated through Chinese visual media intermittently since the time of SARS in the early 2000s, they became far more prevalent after 2013. They include at least three variants. The first involves someone wearing a respiratory mask, into which has been inserted one or more lit cigarette. The second variant is nearly the same, but instead of sticks of lit tobacco, the broken off fibrous ends of cigarettes, the filter tips, are inserted into a mask. The third variant is more a play on consumer personalization: much as marketers have come to offer the public cheap ways to accessorize all manner of functional items—consider the thousands of different mobile phone covers for sale—they are also today peddling respirator masks in innumerable colors and patterns. For the consumer wanting something more than a plain white face covering, options suggested by memers include a mask adorned with a horizontal zipper allowing for easy access to smokes and another one sporting a painted outline of a nose and smiling red lips, gently hanging from which is a painted smoldering cigarette (figure 1).

Should these memes—hereafter what I will call "cigarette masks"—be dismissed as nothing more than senseless dalliances with absurdism? After all, what is more incongruous, more ludicrous than conjoining (a) the respiratory mask, designed to filter out airborne particles with (b) the cigarette, a consumer product that when ignited generates thousands of different chemicals, many of which are immensely pathogenic? In this chapter, I unpack some of the cultural method behind this seemingly comedic madness. Rather than casting them aside as senseless, I look to these cigarette mask memes as a scholarly portal, seeing them as imagistic gateways for understanding air filtration in contemporary China—for surfacing aspects of both its history and emerging affective economy. For one thing, the memes gesture to an uncanny backstory hitherto untold regarding China's current market for home air purifiers. This is a backstory heavily tied to the development of an air filtration product—the cigarette—manufactured across China in far greater numbers today than home purification units. The memes likewise gesture to something significant and unresolved when it comes to air filters: they are double-edged. Residents of the PRC are more and more entangled in environments urging them to enlist filters for toxicological relief. Yet as much as filters may promise a better and less toxic life, and thus seem to be items to be utilized whenever possible, they can also become founts of fraught feelings regarding the inescapability of poisons in a context of profound pollution. Lastly, memes of cigarette masks gesture to a truism too often forgotten in China studies: they remind us that gender politics invariably texture the emergence of any new popular application of science and technology.

Demarcating the Filter

What is the filter? Defining it is not an easy task. Many things might be called "filters" in and outside of China, depending on the projects and disciplines they serve. In their broadest sense, filters are devices that facilitate processes of

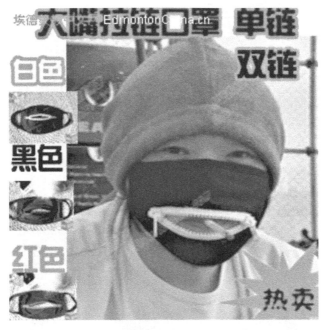

FIGURE 1. Examples of "cigarette mask" memes.

inclusion and exclusion, so they can be as much digital, semiotic, and discursive as they can be mechanical, cellular, and atomic. Social norms can function as filters, as can computer codes, human tissue, and nylon webbing.

Whatever their composition, filters are always political because, when working, they selectively keep things flowing and stop things from getting through. They are also political because what they include, what they exclude, and where they get deployed can benefit some people while imperiling others. Who gets to decide the category of things included and excluded? Who gets to decide how and when the filter is deployed? These are questions foundational to the filter. But there are more. The immediate utility of any filter (keeping something back, letting something through) invariably comes with longer-term effects, some predictable, some less so. Who benefits and who gets hurt by those long term effects? Who gets to decide what's more important, the immediate benefits of any filter or its long-term costs? How can attention to the ostensible short-term benefits of the filter serve to occlude and distract from long-term problems that may precede the filter and become even worse years after its deployment? Who among filter experts—champions/critics of a filter's short-term/long-term benefits or harms—get prioritized?

Filters are likewise political because of their strange relations to binaries. They are all about binaries, keeping a category in or out. But they are also binary busters. After all, they never work 100 percent of the time, and before long, they start to get gummed up and fail ever more to keep tidy divisions. What's more, as we will see later, when placed under the scrutiny of new materialist theory, filters collapse conventional binaries about agency and structure, materiality and meaning.

In this chapter, I discuss a variety of materially tangible filters, ones that are manufactured currently for daily use across China and are ostensibly providing people immediate access to cleaner air. The smallest of the filters that I discuss are no bigger than your pinky finger and typically come twenty to a pack. The largest ones—free-standing air purifiers—come encased in plastic and metal, as small as a toaster, as large as a multi-drawer file cabinet.

Purifiers, Particles, and Puzzling Passions

Air purifies marketed for home use in contemporary China employ a variety of technologies. These can include UV, thermodynamic, and photocatalytic oxidation technologies. Some even use activated charcoal, although this is atypical and is actually an absorption technology. At the center of most air purifiers sold in

China today is an electric fan pushing/pulling air through a replaceable fibrous filter, usually made up of fiberglass or polyester. These materials are layered into lattice formation, differentiating the "filter" from the single layer "sieve." It is in these multilayered lattices where particle retention (filtration) occurs. A large proportion of the purifiers for sale in contemporary China are marketed as possessing a type of filter that by definition is capable of trapping at least 99.97 percent of particles 0.3 microns or larger. They have "HEPA" filters, a technology invented in the United States to remove airborne contaminants for the Manhattan Project, and later trademarked for commercial use.

Not only is the air purifier a product that is commanding more and more showroom space in Chinese department stores; such devices have also surfaced as common topics of my interviews recently regarding people's lived experiences of pollution and domesticity. Consider the words of Ms. Liu:

> It was back in the fall of 2013, when reports of PM2.5 levels first started to jump. We had to do something. So, we bought two Japanese-made air purifiers for our flat. Each cost almost USD 900. A big investment. At first, I was so happy. Those machines . . . they were my heroes.

Relatively quickly, an unseeable object—PM2.5—has become widely known across Asia. It is an object that few had ever heard of in 2005; today it is on the tips of most Chinese urbanites' tongues. "PM" is an abbreviation for *atmospheric particulate matter*, any microscopic object suspended in the atmosphere. Ushered into scientific significance by U.S.-government researchers, initially in preparation for war in the Middle East, particles less than 2.5 micrometers in diameter (approximately thirtieth the average width of a human hair) are today the lingua franca of air pollution experts and menace of public health scholars government officials, and anxious citizens (Jones 2019). Bedeviled by daily tweets of PM2.5 levels from the U.S. Embassy in Beijing starting in 2008, Xi Jinping turned the tables on public relations in late 2012, in the very months between him assuming his titles of general secretary of the Central Committee and president of the PRC. By early 2013, the party-state had nearly 500 air-quality surveillance stations up and running in over seventy cities from which it was releasing daily PM2.5 data (Roberts 2015). Since then, mood swings in those cities have been widely mentioned as inversely tied to PM2.5 reporting. The higher the reported intensity of particulates in the air, the more fearful and dour many people become. How can people protect themselves from all this darkness?

The market response among consumer electronics companies has been predictable. They have introduced a spectrum of *kongqijinghuaqi*: plug-and-play air purifiers. These devices, the companies have pledged, are the savvy urbanite's best remedy to both the material and the affective problems inherent to intensive air pollution. Buy a device and remediation will follow. How men and women have been experiencing any such remediation is an open question, though. For that, let's pick up where we left off with Ms. Liu's reflections.

> Those machines . . . they were my heroes. One year later, though, I was spurning them. When my husband and son are home and the air outside is bad, they typically turn the purifiers on high. But when I'm home alone, I always keep them switched off. It's not because I'm trying to save on replacement cartridges. No, it is because I can't stand seeing the purifiers' LEDs. Those lights—the ones that show the machine is up and running and filtering the air—I hate them. They just remind me how horrible, how unhealthy our lives have become with all this smog and pollution. It's like the air quality apps on my phones. These days everyone uses those apps to decide when and how to go outside. I use them too, but I hate looking at them. The information they give me, as helpful as it is, just brings me down, makes me so depressed.

Why would someone like Ms. Liu turn against what she initially saw as her heroic home air purifier? Is there something about the history of domestic air filtration that would predispose someone like her to be particularly susceptible, more so than her husband or son, to what we might call filtration disenchantment? Are we hearing an undercurrent of matriarchal critique running through Liu's comments? That, in turn, raises the issue, what do we know of the gender coding of home air filtration in China?

Blue Erotics

To the cigarette mask once more. Notably, few images that I have found of the masks feature a female user. It is almost always a man donning them. That makes sense given the stark disparity in cigarette smoking prevalence among the sexes today: in China currently over half of men are daily cigarette smokers, whereas less than 3 percent of women are. It also makes sense if we read cigarette masks as communicating issues about gender and the inescapability of toxins in a context of profound pollution.

The cigarette mask is a merger of timeworn scripts at the intersection of gender and material culture: a heavily male-coded tobacco product (the cigarette) and a decidedly female-coded filtration technology (respiratory mask). Whereas women in China have for decades been encouraged to demonstrate responsible femininity in the face of aerosolized particulates by wearing a respiratory mask in public, most men had been the recipients of a very different message, save for the few involved in specialized medicine or who, in 2003, were directly involved with the SARS epidemic. Until the arrival of the first *kongqimori* (airpocolypse/airmageddon) in February 2013, real men (*nanzihan*) were typically configured as those willing to risk and sacrifice themselves when it comes to aerosolized toxins, whether associated with the extremes of mining, industrial manufacturing, or warfare. The ongoing currency of that logic was communicated by Xi Jinping as recently as 2014. In February of that year, hours after government air quality officials had announced the country's first PM2.5 Orange Alert, President Xi took a well-publicized stroll through a central Beijing neighborhood. "Breathing Together, Sharing the Fate" was the Xinhua headline describing Xi's walk. Stunts like this have not been lost on sycophantic foreigners seeking to curry favor with Chinese leaders for greater market access, as we saw in March 2016, when Mark Zuckerberg did a highly publicized "smog jog" (sans mask) through central Beijing when PM2.5 numbers were topping 300.

Given this symbolism, it is no surprise that some men have come to poke fun at facial filtration devices, in the case of cigarette mask memes, literally skewering them with cigarettes. Also, it is no surprise that marketers have turned to a highly gendered palette to insinuate the air purifier into a nervous public. Marketers of purifiers have painted a portrait that promises consumers a technological sublime. The respiratory sanctuary they advertise typically appears in light pastels and modern architecture. This is a heteronormative, pronatalist paradise populated by children and/or young attractive women relaxing in casual garb—with husband/father nearby or off at work. The purifier stands guard, delivering an erotics of health, happiness, and racialized mobility. Air purifiers' branding and pricing in China today depends heavily on purported national provenance, with the most expensive purifiers coming from Northern Europe. Chinese-language marketing copy for purifiers made by companies like Blue Air of Sweden promise buyers not just clean air but also the bonus of a superior lifestyle demarcated by the accrual of prototypical Northern European skin tones and facial features (figure 2).

Confronted with snow-white marketing and darkening skylines, people have been buying up purifiers at a feverish pace. Brightly lit department store showroom display a wide variety across price points. The internet is rife with advertisements and consumer reviews. From taxi drivers to office workers, innumerable people have told me about purchasing purifiers. A friend of mine told me that she made a discount bulk order of purifiers branded by her employer, the China Center for Disease Control and Prevention (China CDC), gifting over ten of them to her parents, aunts, uncles, and friends. A colleague who lives in a new gated compound was happy to show me the European branded devices he purchased for his wife and infant daughter. I have met just as many men as women who have been the actual purchasers of the machines, not just the users, and nearly all

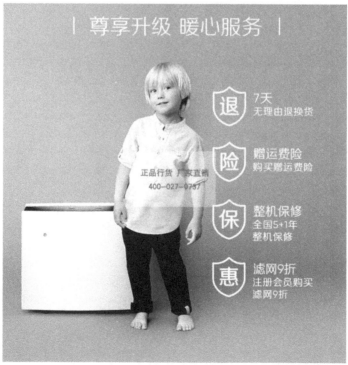

FIGURE 2. Marketing materials issued by the Sweden-based Blueair Company's China division.

of them have been eager to share with me their feelings of initial relief after bringing home their purchase. And no shortage of purifier envy exists, with rumors
running rampant of how the government elite have dotted their workplaces and
homes with purifiers costing over USD 1,500 each (Li Yang 2014).

Filtration Disenchantment

Marketers and manufacturers of purifiers have not been without their detractors.
Editorial cartoonists have helped visualize the most common lines of critique. One
line, as shown in figure 3, has been that purifier merchants are cashing in at the
intersection of pollution and inequality, exploiting social stratification and treating as disposable those who have no choice but to labor outdoors in the pollution, including the poor who are expected to trade their health to haul the trappings of the good life for the middle class to and from stores and trash heaps. The
second line of critique is that purifiers are technologically unreliable, not necessarily able to deliver on their promises of durability and rates of filtration.

These lines of criticism invite affective response, as much as does the trope of
the technological sublime used by advertisers. Emotions that viewers are prompted
to feel when encountering the critiques seem to be everything from doubt, guilt,
and perhaps anger. But these are not the feelings that women like Ms. Liu have
described to me. Rather their sentiments have leaned toward that of fear, endangerment, captivity, desperation, and hopelessness.

Since first hearing Ms. Liu's sentiments, I have wondered why she more than
her husband and son might be prone to this type of filtration disenchantment?
Are some women more than some men susceptible to feelings of fear, endangerment, captivity, desperation, and hopelessness when exposed to air purifiers? If
so, why? Or should this puzzle be looked at from an opposing angle? Might something be keeping men from being predisposed to those same emotional responses? Ms. Liu's sentiments certainly seem like reasonable ones when confronted in the intimate spaces of home by messaging of environmental ruination
and uncertain technologies of remedy. Are some men somehow or another in-
oculated from those feelings? And if so, how?

Two years ago, when I first started collecting accounts of Chinese homeowners buying air purification machines and their emotional responses to those purchases, I felt certain that I was venturing into a domain of material culture fundamentally new to a country that I have been studying for nearly three decades.
After all, during the 1980s, when I first began spending lengthy periods in China,
what retailer there sold any devices singularly designed for purifying unseeable
particulate matter? Cloth face masks could be found for sale in some settings, to

FIGURE 3. Cartoon of an air purifier deliveryman.

Source: http://src.house.sina.com.cn/imp/imp/deal/ca/f6/2/d526283672e09a611d45bec95bc_p1_mk1.jpg.

be sure, but they were typically sold as things to use episodically to defend against heavy and visible dust, like that experienced during Beijing's spring sand storms. The expectation that one will buy and use fibrous filters to cope indefinitely with aerosolized objects so small that they can be viewed only under a microscope is altogether new to China, right?

Founts of Filtered Life

Again to the cigarette mask. In the remaining pages of this chapter, I consider an uncanny backstory to the insinuation of filtration technology into people's lives. Rather than trafficking in an academic narrative of unparalleled contemporary novelty, I consider current day adoption of and gendered responses to air purifiers in China in dialogue with an earlier moment in the psychic life of air filtration, a historical epoch that predates terms like *airpocalypse* or *airmageddon*. This archival work requires us to ruminate less on the contemporary home air purifier per se than on one element in the ostensibly comedic comingling of respiratory masks and machine-rolled tobacco products. What I chronicle below is a period in the late twentieth century, when manufacturers flooded Chinese shops with a new category of air purifier, initially just a few million a year, then tens of billions, and before long trillions. Today, nearly a third of all Chinese citizens use this specific category of filter on a daily basis. The origin story that I recount is of *guoluzui*—that is, filter-tipped cigarettes—how they came to flourish, and how they came to texture some feelings about filtration, airborne toxins, and breathing in the unseeable and unescapable.

Consideration of this earlier and overlooked moment of filtration mania emboldens the following arguments. First, the current adoption of air purification technologies by homeowners in China, rather than being unprecedented, is of a piece with past cultural scripts and events. Second, the image of someone smoking through a filtration mask, rather than being absurdist, makes ample sense—in fact it might make more sense than someone just using a mask without a cigarette attached. Third, air filtration in Chinese contexts has been generative of a way of being human—what I call "filtered life"—that has been little understood, even though it has been unfolding for nearly half a century. This is a materially mediated form of existence (which all existence always is) wherein the physicality of the air filter becomes something that allows people to experience, cope with, and shape changing forms of politics and ethics about what it means to be human in a time of heightened anxiety over aerosolized ruination. And fourth, understanding how such filtered life has unfolded, through the innovations in the design of cigarettes, casts light on a gender-inflected critique of China's environmental ruin.

Filtration Flimflam

What do we know about the history of filter-tripped cigarettes (FTCs)? The story begins outside of Chinese-speaking contexts. European and American companies began producing FTCs in the 1930s, with the first commercially successful brand being Viceroy, released in 1936 by the Kentucky-based outfit Brown and Williamson, under a slogan reading a "Safer Smoke for Any Throat" (Proctor 2011, 343). In the United States, filter-tipped cigarettes remained largely a novelty product until the 1950s, often advertised as delivering vague therapeutic effects. After World War II, with new scientific findings published regarding tobacco's harm, U.S. tobacco firms launched filter-tipped brands like Winston and Kent, and their scientists made bold claims including that air breathed through those ignited products is "several times cleaner than the air you normally breathe in an average American city" (Proctor 2011, 340). Smokers bought these cigarettes and their hype, "believing harmful chemicals were being removed" (Proctor 2011, 345).

Even more bought them after 1964, when the U.S. Surgeon General announced that smoking causes lung cancer. By 1980, the conversion to filter-tipped products in the United States and much of Western Europe was nearly complete, and the industry there was going strong, with filtration having become the centerpiece of the its fabulously successful risk-reduction deception, targeting both men and women alike. The deception of "low tar" and "light" cigarettes had been fully folded into FTC branding, advertised under innumerable labels carefully configured to gender, class, and social demographics. The most successful label in this flimflam was Marlboro, reintroduced in the 1950s by Philip Morris with "new and improved" filter tips—flimflam because filter-tips provide the public no protection from the vast majority of deadly chemicals that ignited tobacco releases, a fact secretly known within the North American industry since the mid-twentieth century; flimflam because FTCs are actually more dangerous than the products that they replace, given that they compel customers to smoke more sticks per day and suck smoke deeper into the lung in order to achieve a desired nicotine dosage (Pauly et al. 2002; Proctor 2011; Stevenson and Proctor 2008).

Pandering to Party Patriarchs

The filter-tipped cigarette is a slightly newer phenomenon in China. Its manufacture did not occur until after the Communist Party had come to power in 1949 and had nationalized the country's cigarette supply chain. Initial production of FTCs in the People's Republic was prompted by the rising prevalence of filter-tipped cigarettes elsewhere but worked off a decidedly different emotional and gendered palette. Rather than exploiting anxieties among male and female smokers about the health risks of tobacco, the introduction of FTCs capitalized on another emotional terrain: mainland men's need for political affirmation. And the pandering started at the top. In the late 1950s, the first line of FTCs were produced to celebrate the tenth anniversary of the revolution and its leadership. The Shanghai Tobacco Company won authorization from Beijing to have a Hong Kong front company buy a filter-making machine from the German Hauni Corporation. Shanghai Tobacco used the machine together with filter tips acquired from a Japanese supplier to create a limited edition of its Chunghwa ("China") cigarettes. Until the 1980s, Chunghwas were rarely for sale, reserved for exclusive use by party elites. The Shanghai Tobacco Factory was not alone in creating and reserving superbly formulated brands of cigarettes for Communist elites. The Kunming Cigarette Factory, for instance, launched its now flagship brand, Yunyan, for exclusive use by PRC luminaries. Indeed, there was something of an arms race that emerged in Mao's China, with cigarette factories vying for recognition of their ability to balance production of mass-market labels alongside a highly exclusive small-batch brand loyally favored by top officials. Not only was Chunghwa the first filtered edition of any of these small-batch cigarette offerings, but it was delivered to none other than Mao and a small group of his brethren at the apex of the polity. The gift proffered was pure patriotism: the finest tobacco that China grew, machine rolled by government factories, and replete with the most modern of smoking device—the filter tip. What was the payoff? To a certain degree, the prize was the affect of an elevated male psyche, enjoyed by both a small coterie of male government elites and the men who ran the cigarette factories fabricating the newly formulated filter-tipped sticks.

In China filtration-free cigarettes still remained the norm until the 1980s. With scant information about the health consequences of tobacco circulating in the country during the Maoist years, adopting market-wide risk-reduction techniques including filter tips was a low priority for the industry. Also, few manufacturers had enjoyed access to the necessary cash, connections, and political cover to buy "foreign" equipment and materiel to roll out FTCs. Even at the prestigious Shanghai Tobacco Factory, only one out of every 100 cigarettes coming off the production line as late as 1972 was filter-tipped (Shanghai Tobacco Gazetteer n.d.). What changed the tide, what triggered China's state-owned cigarette industry to convert nearly all its production over to filtered cigarettes, were the political and fiscal priorities let loose by the cessation of the Cultural Revolution (1966–1976). Exhausted and demoralized by years of Maoist ardor, the Chinese Communist Party (CCP) set a new course for the People's Republic in the 1970s, one that prioritized a more technocratic, market-driven definition of how the party-state and citizenry should achieve modernity. One of the earliest commercial beneficiaries of this change was the country's tobacco industry.

An Audience Susceptible to an Intoxicating Symbolic Brew

Across the country as Maoist zeal faded, local governments realized that to win cash infusions from Beijing, they had two tobacco cards to play. One was the ramshackle cigarette factories that they already controlled; the other, the prestige of filtration. An argument that provincial and municipal governments made to the Ministry of Finance was that capital investments in the production of more modern cigarettes would set off a chain reaction. It would stimulate cigarette sales, which would shore up local tax revenues, which would loosen up money for other local market liberalizations. Central to this argument was that old cigarette factories should be overhauled and new ones built with all the equipment needed to produce a technologically distinctive category of tobacco product, the filter-tipped cigarette. Officials from Yunnan province were especially quick to rally Beijing's support along these lines, and officials from other regions were right behind. Why was the filter-tipped cigarette identified in the 1970s as an especially promising approach to light a fire under purse-holders in Beijing and cigarette smokers across the country? The potential for making a supposedly safer cigarette was not meaningless at this juncture. Industry leaders knew a problem lay on the horizon given their monitoring of growing scientific evidence published outside of China regarding tobacco's toxicity. The overwhelming rationale for moving into large-scale FTC fabrication, however, was far more about the promise of this product category delivering an intoxicating symbolic brew of science, Western modernity, and prestige to a highly susceptible audience hankering for all things "modern," all things "Western." FTCs had already become the industry standard outside of China—well over 80 percent of cigarettes manufactured by companies based in Japan, Britain, Germany, and the United States were filter-tipped by 1975. King among foreigners was Marlboro, and China needed to catch up. No less significant was that FTCs carried all the prestige of the CCP elites. Domestic media had for years regularly featured images of Communist VIPs enjoying filter-tipped cigarettes. It was a public secret that Mao's favorite brand was the filtration trailblazer, Chunghwa.

Susceptibility to FTCs had much to do with gender as well. If you were a woman, you were unlikely to have been a tobacco smoker after 1949 because of how thoroughly the new party-state vilified female smoking as an anathema to female propriety and Maoist ethics of nation building (Benedict 2011, 199–239). By contrast, being male during this period meant you had a much higher likelihood of feeling encouraged to experiment with cigarettes as a teen, smoke them daily in public as an adult, exchange them with others (using the symbolism of cigarette brands to project your personality), and modulate your emotions through the nicotine these products delivered (Kohrman 2018, 1–33; Benedict 2018, 94–95). So, it was men, much more so than women, who invariably played an active part in the uptake of filter-tipped products from the 1970s forward. What's more, during the latter half of the twentieth century, being categorized male also meant that you were far more likely to be drawn into something else than if you were categorized female. An unspoken screen for recruitment into leadership roles of the party-state—up and down the bureaucracy—was (and remains) gender. Compared to their female comrades, male members of the party have been far more likely to make it through the selection processes of promotion into positions of higher authority. And, once a man had risen to that august platform, one of the many trappings of exalted party status until the 1980s was access to and the smoking of the finest brands of cigarettes that the Chinese tobacco industry had to offer. Manufacturers of newly released mass-market brands of FTCs were not shy about trafficking on that and other symbolism of upward mobility. And their unspoken poster boy was none other than Mao's replacement. Come the 1980s, yet another public secret was that Deng Xiaoping—who had catapulted from denounced internal exile during the Cultural Revolution to preeminent leader after Mao's death in 1976—smoked a newly filter-tipped label. Deng smoked a version of Pandas with extra-long filters, which had been designed exclusively for his use.

Male smoking of FTCs was not restricted to CCP members, however. Men of all backgrounds adopted them from the 1970s forward. The tobacco industry facilitated adoption through astute attention to branding, pricing, and consumer behavior. Local factories created hundreds of new labels, symbolically targeting men based on region, occupation, and educational achievement. At first, the words denoting *filtration* were given prominence on these cigarette labels (figure 4).

As years went by and FTCs became the industry standard, words denoting filtration began to fall away in marketing copy. Once the technology had been fully normalized in the material world of smokers, it no longer needed to be called out on packs. Meanwhile, at the end of the twentieth century, another marketing strategy supercharged FTC adoption. During the very years that FTCs became the industry standard, factories expanded the price points between their most expensive labels and their cheapest ones, typically creating ladders within their most successful brands. Just as there are many different labels of Marlboros sold today around the world (Reds, Golds, 27s, Edge, and so on), cigarette factories in China have come to create different labels within a brand. A key contrast in this foreign analogy is pricing. Whereas little difference exists between the price of the most expensive label of Marlboro and the cheapest, during the last years of the twentieth century as much as a tenfold price span opened up within ladders of Chinese FTCs. Like branding and FTC labeling, price differentiation helped fuel FTC adoption by lighting a fire under consumer habits long common to the Chinese cigarette market: homosocial gifting of cigarettes between male smokers and the use of cigarettes as symbolic semaphore of social status (Hessler 2010, 232; Kohrman 2008). What label of FTC you smoke, what label of FTC you gift, became vitally important in the lives of many men, as echoed by this Kunming resident:

> MK: Do you remember when you could begin to buy filter-tipped cigarettes?
>
> MA LUZI: Sure, I started smoking in the 1980s. I experimented a little when I was in high school. But I really had going when I had my first

FIGURE 4. Three pack designs from the late twentieth century, each with words denoting filtration called out by an arrow. The brand name of the middle pack is "Strongman."

FIGURE 4. Continued.

FIGURE 4. Continued.

job. It felt like you had to smoke to succeed. My first boss smoked. He would give me a cigarette to smoke. Of course I had to smoke it. And the expectation was that I would have cigarettes of my own to give him. Cigarette gifting, that's how it all got started. He smoked a filter-tipped brand. So, it would have been insulting if I gave him cigarettes without a filter. At first, I smoked a cheaper brand when I was alone and only gifted filter-tipped cigarettes. But as time went by, I made a little more money and so I wanted to smoke something better, so I changed brands to one that had a filter. We used to talk about the filter back then. Not so much now. Now it's all about taste, the brand's home province, how much it costs. How much difference is there between all these brands, I don't know. Probably not much. In the end, after you've smoked them, what's left is all the same, the used-up filter.

MK: Is it important that all the cigarettes for sale come with filters?

MA LUZI: I guess so. Better that they have them than not. I'll take all the protection that I can get. But in one way, they have become a problem. They are what's left over when I smoke and my wife doesn't want me to smoke, especially at home. I used to smoke at home but not anymore. She gets angry when she comes across any sign that I've been smoking at home. If she comes home and smells tobacco smoke or finds cigarette filters in the garbage, she gets angry.

By the early 2000s, it was not just the proliferation of filter-tipping of cigarettes that seemed unstoppable. Something else regarding the cigarette was inexorably on the rise: a recoding having to do with danger, the biosciences, and gender. Pack design did not escape this recoding. In 1991, the National's People's Congress began requiring that cigarette packs sold in China have printed on their sides tar levels and the words *xiyanyouhaijiankang* "smoking is harmful to health"). In the years that followed, Beijing took further steps, requiring cigarette makers to also disclose on the side of packs how much, when consumed, each generates in milligrams of nicotine. And, a few years later, it has required that carbon monoxide be disclosed. Left vague for the consumer throughout and ever since these disclosures have been mandated is whether the numbers listed on the sides of packs are indicative of what gets released before or after smoke passes through a cigarette's filter. Many smokers have told me over the years that they have understood disclosures as pre-filtration readings. As Mr. Chen Bolai once told me in Kunming, "Most of those bad things get removed by the filter. That's why they put the filter on the cigarette. That's why I'd never smoke an unfiltered cigarette."

Not in My Home: Public Health Radicalizes Wives

Around the same time that warnings and disclosures began appearing on the side of packs, public health advocates across China began using gender in a new way to communicate the perils of tobacco smoke. Health advocates in the 1990s portrayed cigarette smoke as dangerous for everyone but especially so for women. This was not an altogether new tactic for tobacco control in China. Women had been heavily discouraged from smoking cigarettes by anti-imperialist critics prior to 1949—starting in the 1920s many women had taken up cigarette smoking—and after Mao came to power, the CCP even more intensely damned the practice of women smoking as highly transgressive for their social standing and dangerous to their feminine propriety. Come the 1990s, for women, the erstwhile symbolic dangers of tobacco smoke were fused to a newly categorized biological danger: "environmental pollution" (*huanjingwuran*) emanating from the lit tobacco in other people's hands. Women were instructed to avoid tobacco smoke in all forms, especially if they were pregnant. By the 2000s, this was being emphasized anew in the language of "second-hand" smoke.

Significantly, this fusion of symbolic and biological toxicity allowed women to exercise new authority over their male family members. Often proportionate to household educational level, men started to feel increasingly browbeaten by mothers, wives, and daughters for smoking cigarettes and increasingly pressured to step outside whenever feeling the urge to light up at home. By 2000s, among the more educated, the home had become something of a tobacco battle ground. More and more women were no longer willing to sit idly by as their men contaminated their homes. As much as the filter-tipping of cigarettes had placated many men's anxieties, allowing them to continue servicing their powerful nicotine addictions, women were being radicalized to cleanse their homes of tobacco smoke. Women enlisted into this campaign of household filtration typically gave little credence to spongy white cellulose acetate filters on the ends of cigarettes. They provided them no protection from second-hand smoke or the latest specter, third-hand smoke (residual chemicals left on indoor surfaces by tobacco smoke). Instead, these filters became for some women ugly domestic detritus, reminders discarded in their homes of defilement and discord, as explained by two Beijingers:

> Ms. Zhang Ting: No smoking in the house. That's what I tell my old man. No smoking. After we married and moved in together, he began smoking more and more. Every few days, it would be my job to empty his ashtrays. Now that I know how dangerous smoking is, I have become much braver in telling him what he can and can't do. No more smoking in the house. No more dirty butts piling up on our tea table. Second-hand smoke, third-hand smoke. I know all about that.
>
> Ms. Shen Xiaowen: My husband tells me that I shouldn't ban him from smoking in the house because they all have filters and he buys more expensive cigarettes. But I know they are still dangerous. And women

are even more at risk than men. I've heard of many women who've contracted lung cancer and died. They never smoked. So how did they get lung cancer? Being around so many cigarettes. More expensive cigarettes, filtered cigarettes . . . it doesn't matter. I don't believe any of that. But [sigh] unless I'm willing to get divorced, there is nothing that I can do.

Feminist Filtration of Science and Technology Studies

Back to the comedic—this time outside of China. In a 1999 essay, Bruno Latour begins with a cartoon strip by the Argentinian Joaquin Salvador Lavado (aka Quino). The strip features a domestic scene occupied by father, daughter, wife, and cigarettes. It begins with the father contentedly at home in an armchair puffing on a cigarette, his daughter, Mafalda, watching fretfully. The strip ends with the father using a pair of scissors to chop up his remaining cigarettes, mother watching with bemusement. Here's how Latour describes the cartoon's narrative arc:

> "What are you doing?" [Mafalda the daughter] asks in the first scene. "As you can see, I'm smoking," responds her father unwarily. "Oh," Mafalda remarks in passing, "I thought the cigarette was smoking you." Panic. Whereas he thought of himself as an untroubled father, comfortably seated in his armchair after a hard day at the office, his daughter saw him as an unbearable monster: a cigarette grabbing a man to have itself smoked in a big cloud of tar and nicotine; the father as an appendage, an instrument, an extension of the cigarette, the father becoming cigarette to the cigarette. (1999, 21)

Latour goes on to use this cartoon as a touchstone for thinking about how the social sciences have been forced into a binary about the relative agency enjoyed by people and objects. Since then, we have seen anthropology drawn into a productive but at times paralyzing discussion of this binary. Like other anthropologists who study China and Asia, I "would prefer to avoid the rather arid debate about whether objects have agency" (Herzfeld 2015, 19; also see Kipnis 2015). Instead, my aim is to pursue a different analysis of the arid, one centered around gender and affect as much as it is home air filtration. As Haraway pointed out as early as 1997, baked into Latour's science studies at the turn of the millennium was a disregard for social categories like gender. One sees this clearly in Latour's analysis of the Malfalda cartoon. In his rush to puzzle over thingness, Latour highlights one aspect of the lit cigarette—its prompting Mafalda to question her father's agency—but altogether ignores another aspect looming large in the strip's last frame. There, from a doorframe, Malfalda's mother looks across at us fretfully (figure 5). The look is one of utter dismay at the father despoiling the tidiness of their living room with chopped up cigarettes while she works two jobs: (a) comporting herself elegantly, standing upright in a miniskirt, slender legs exposed down to the ground and (b) tending to domestic chores, transporting a stack of folded garments.

Feminist scholars have reminded us that when analyzing new technologies, we should use care adopting analytics for understanding ontology. Whether drawing on science studies approaches, Marx's commodity fetishism, or other perspectives, we should not let practices of sociopolitical difference, including ones pertaining to gender, drift out of our purview, for always "objects of technoscience are forged and branded in the crucible of specific, located practices" (Haraway 1997, 35). Sara Ahmed (2010) has moreover encouraged us not to be content simply falling back on Appadurai's general call, following on Kopytoff (1986), to collect the "cultural biography of things," how objects "move through different hands, contexts, and uses" (Appadurai 1986, 34). She prods us instead to pay keen attention to the intersection of bodies, space, and affect. It is there, at that phenomenological intersection, she argues, where gender and objects are mutually naturalized and contested. She suggests that in order to better understand such naturalization and contestation scholars need to provide richer descriptions, which tack back and forth between object and gender, depicting how each is "experienced from the points of view of those who share the space of its dwelling" (244).

Among the men and women I have met in China recently dwelling with air purifiers, many have for years also dwelt alongside filter-tipped cigarettes. The domestic scene for the three (men, women, and cigarettes) was fraught before the introduction of purifiers in the 2010s. The fault lines of affect have become even more complex in many cases since then. To recap, over the last thirty years, the proliferation in China of filter-tipped cigarettes and pack disclosures regarding nicotine, tar, and carbon monoxide were quickly followed by an intensification of tobacco-control discourse, often playing up women's unique precarity. Many men bought into the tobacco industry's filtration rhetoric while ignoring any tobacco-control discourse. Other men consumed both, coming to feel conflicted about their ongoing nicotine addictions, yet unable to quit (Kohrman 2004). Among women, few were using tobacco before the rise of FTCs, and few do today. For them, the arrival of cigarette filters typically became meaningless addendums to a product. Some have told me that they still like the smell of cigarettes, a reminder of meaningful ties with men in their lives—father, grandfather, husband. But for many, particularly the better educated, FTCs and the detritus of "butts"

FIGURE 5. Cell six of cartoon by Quino, which originally appeared in his *Le Club de Mafalda*, no. 10 (1986), p. 22.

came to be synonymous with airborne dangers that, through tense negotiations with their menfolk, they had either successfully purged from their homes or were constant smelly reminders of how they were stuck with men who were more inclined to place personal pleasure over responsible care of others. To be sure, by 2010, many households had become even more stressful dwellings as a consequence of FTCs.

Lauren Berlant, in *Cruel Optimism*, investigates affective stories about the "dissolution of optimistic objects/scenarios that had once held the space open for the good-life fantasy" and tracks dramas of adjustment from relations propping up optimism to ones that are cruel (2011, 3). It is interesting to note the timing of the dual dissemination of FTCs and highly gendered second-hand smoke discourse in China and how it coincided with a surging practice of radical domestic disenchantment across the country—that being divorce. Divorce rates doubled between 1979 and 1986, and then more than tripled between 1990 and 2011 (Lu and Wang 2014; Platte 1988). Of course, it would be foolish to attribute surging divorce rates to cigarettes and their filters. But if nothing else, rising rates of divorce point to the degree to which relationships between husbands and wives have become increasingly subject to disagreement and dissolution, and how fantasies of the good life that are so foundational to conjugation have been under tremendous strain.

It is into that context that large-scale marketing of home air purifiers arrived in 2013. Well after divorce rates had begun to surge, PM2.5 became a bane of many people's existence, and air purifiers became a reputed remedy. Since then, I have heard numerous stories of men petitioning their spouses, long irritated with cigarettes in the home, to consider the purchase of air purifier as achieving two functions: cleaning the home of toxins introduced by outside air pollution and cleaning the home of toxins introduced by cigarette smoke. This has not been by happenstance.

First, when Beijing introduced nationwide daily PM2.5 disclosures in early 2013, home air purifiers were hardly new technologies to some smokers. In the early 2000s, outfits like Shenzhen's Shenrui Corporation (figure 6) had begun to convert industrial air filtration technologies into some of the first home air purifiers, which they marketed as "Tobacco Fog Purifiers," capable of "thoroughly purifying second-hand smoke!" Such claims were thereafter given broad airing through some of the country's most prestigious scientific media (Zhang 2007).

Second, when the terms *airpocolypse* and *airmageddon* entered in the Chinese zeitgeist and home air purifiers suddenly became must-have purchases for many urbanites, they were immediately advertised as capable of filtering out pollutants generated from a variety of sources including automobiles, kitchen stoves, and cigarettes. Today, online advertisements for home air purifiers continue to include second-hand tobacco smoke in their lists of what their devices effectively filtrate.

The point that I'm making here is that husbands and wives were prompted by marketers to experience the adoption of air purification technology after 2013 as an extension of existing tensions in their household. Some husbands saw these machines as "green lights" to continue or recommence smoking at homes. The machines promised a new filtration fantasy enabling them to once again feel that they were pursuing the good life, enjoying cigarettes in responsible ways. To wives of these men, it was all too easy to feel, before long, that the machines were Trojan Horses. Consider the experience of this Beijing husband.

> Liu Zhi: We bought an air purifier because the PM2.5 levels had become so high. That was the right thing to do. But once I had the machine in the house and had a chance to read through all the literature that came with it, I realized that I could smoke cigarettes again in the house. The machines purify the air of pollutants, and that includes

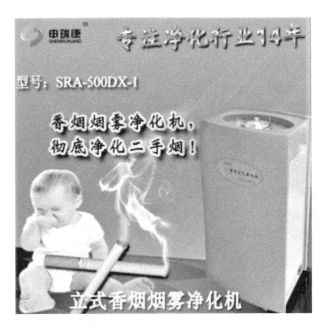

FIGURE 6. Ad for Shenzhen's Shenrui Corporation's "Tobacco Fog Purifier."

Source: https://detail.1688.com/pic/522062794372.html?spm=0.0.0.0.JPsl3S.

what's in cigarette smoke. That fact just made my wife angry. She thought she'd convinced me to no longer smoke at home. With the purchase of the air purifier, she came to feel that I was trying to trick her. I told her that the purifier wasn't a trick. I was just being realistic and practical. I have tried to quit. I don't really want to smoke if I could stop, but I can't. And I hate having to always go outside to smoke. The air purifier is a safe solution.

Of course, the idea of a safe technological remediation to aerosolized toxins is exactly the language that has underwritten tobacco industry rhetoric around the world for years, exactly the logic that underwrote the filter-tipping of cigarettes elsewhere in the world and that came to help FTCs' popularization in China.

But what about the Chinese households wherein the rhetoric of the "safer cigarette"—tipped by filter—has not been materially present over the years, households that have been long free of cigarette smokers? This brings us back to Ms. Liu, the interviewee with whom I began this chapter. Ms. Liu's husband is an engineering professor and has not smoked since college, and thus has not been pressing her to let him smoke cigarettes anew in their household just because they now have two expensive air purifiers. How can we understand her disenchantment with those air purifiers, her preference to turn them off?

> Ms. Liu: Yes, normally, I'm not the one to turn the machines on and I often turn them off when everyone else leaves. As I said, I was so happy with them at first. I believed in them. But they just make me feel like a caged bird now. Am I to just stay home all the time? What solution is that? It just makes me so mad. It just makes me so sad. The situation that we are in now. That I have to sit in my home and worry about all the pollution flowing into it. I didn't make the decision to build all those factories, to sell all those cars. Government officials and their friends did that. They are the ones who have promised us a bright future, but made all our lives so bad. And frankly, I'm not sure I believe that purifiers make much of a difference. I have read lots of skeptical information online and in the newspapers which say that the machines don't really help much. That's just another difference between me and my husband—there are so many—which is kind of strange because he's the engineer.
>
> Prof. Zhang (Ms. Liu's husband): Of course, I turn the machines on. We have to do whatever we can to find modern technologies to fix modern problems. We can't go backward. There are technological solutions for this big problem, I'm sure of it. The most important technological solutions will have to be made at the level of all this pollutions' production. Clean, green energy. That's the future. But until then, this is the best solution. Whatever purifiers and filters everyday people can get access to . . . they should use them. I didn't wear a face mask before, but I do now. Whatever filters everyday people can get access to . . . they should use them.

I read Ms. Liu and Prof. Zhang's words as part of a gendered politics of pollution emergent in China today. This is a gendered politics that is of a piece with Chai Jing's influential, self-financed 2015 documentary film, *Under the Dome*. Chai Jing's is a feminist-inflected environmentalism, one that sees much of the decision-making that has led to the defiling of skies across the PRC as generated by misguided, criminal, corrupt, and mostly male government-corporate profit-seeking. This is a gendered politics of affect, born of the most intimate of spaces, the home, and the most public of spaces, the nation. This is a gendered politics wherein citizens like Professor Zhang continue to be seduced by or addicted to the fantasy of a good life animated by technological innovation and remediation. This is a gendered politics in which, for many men, types of filtration are familiar and worthwhile ways for them and their families to keep plowing ahead. Lastly, this is a gendered politics wherein the comedic of the cigarette mask is at once a sardonic nod to the impossibility of people's circumstances and an assertion by such jokesters that filtration (however cobbled together) is the best way for many to endure.

References

Ahmed, Sara. 2010. "Orientations Matter." In *New Materialisms: Ontology, Agency, and Politics*, edited by Diana Coole and Samantha Frost, 234–256. Durham, NC: Duke University Press.

Appadurai, Arjun. 1986. *The Social Life of Things: Commodities in Cultural Perspective.* Cambridge, UK: Cambridge University Press.

Beck, Urlick. 1992. *Risk Society: Toward a New Modernity.* London: Sage.

Beck, Urlick. 1999. *World Risk Society.* Cambridge: Polity Press.

Benedict, Carol. 2011. *Golden-Silk Smoke: A History of Tobacco in China, 1550-2010.* Berkeley: University of California Press.

Benedict, Carol. 2018. "Bourgeois Decadence or Proletarian Pleasure? The Visual Culture of Male Smoking in China aross the 1949 Divide." In *Poisonous Pandas: Chinese Cigarette Manufacturing in Critical Historical Perspectives,* edited by Matthew Kohrman et al., 95–132. Stanford: Stanford University Press.

Berlant, Lauren. 2011. *Cruel Optimism.* Durham, NC: Duke University Press.

Haraway, Donna. 1997. Modest_Witness@Second_Millennium.FemaleMan_Meets_ OncoMouse. New York: Routledge.

Herzfeld, Michael. 2015. "Anthropology and the Inchoate Intimacies of Power." *American Ethnologist* 42(1): 18–32.

Hessler, Peter. 2010. *Country Driving: A Journey through China from Farm to Factory.* New York: Harper.

Jones, David S. 2019. "Air Pollution in India: Why It Matters That It Is Not a New Problem," paper presented at the workshop, "Sensation, Perception, and Policy Intervention: Air Pollution in China," Harvard University.

Kay, Samuel, Bo Zhao, and Daniel Sui. 2015. "Can Social Media Clear the Air? A Case Study of the Air Pollution Problem in Chinese Cities." *Professional Geographer* 67(3): 351–363.

Kipnis, Andrew. 2015. "Agency between Humanism and Posthumanism: Latour and His Opponents." *HAU: Journal of Ethnographic Theory* 5(2): 43–58.

Kohrman, Matthew. 2004. "Should I Quit? Tobacco, Fraught Identity, and the Risks of Governmentality in Urban China." *Urban Anthropology* 33(2–4): 211–245.

Kohrman, Matthew. 2008. "Smoking among Doctors: Governmentality, Embodiment, and the Diversion of Blame in Contemporary China." *Medical Anthropology* 27(1): 9–42.

Kohrman, Matthew. 2018. "Introduction." In *Poisonous Pandas: Chinese Cigarette Manufacturing in Critical Historical Perspectives,* edited by Matthew Kohrman et al., 1–33. Stanford: Stanford University Press.

Kopytoff, Igor, 1986. "The Cultural Biography of Things: Commoditization as Process." In *The Social Life of Things: Commodities in Cultural Perspective.* Edited by Arjun Appadurai, 64–91. Cambridge: Cambridge University Press.

Latour, Bruno. 1999. "Factures/Fractures: From the Concept of Network to the Concept of Attachment." *RES: Anthropology and Aesthetics* 36(Autumn): 20–31.

Li, Yang. 2014. "Zhongnanhai li shenme kongqi jinghuaqi? Yuanda TA2000" [What Air Purifiers Are Inside Zhongnanhai? Broad TA2000]. *Paopaowang*, September 16.

Lora-Wainwright, Anna. 2010. "An Anthropology of 'Cancer Villages': Villagers' Perspectives and the Politics of Responsibility." *Journal of Contemporary China* 19(63): 79–99.

Lu, Jiehua, and Xiaofei Wang. 2014. "Changing Patterns of Marriage and Divorce in Today's China." In *Analysing China's Population,* edited by Isabelle Attane and Baochang Gu, 37–49. Cham: Springer Nature Switzerland.

Magnier, Mark, and Lilian Lin. 2016. "April Fools' Day Is 'Inconsistent with Core Socialist Values,' Chinese News Agency Says." *Wall Street Journal*, April 1.

Pauly, J. L., A. B. Mepani, J. D. Lesses, K. M. Cummings, and R. J. Streck. 2002. "Cigarettes with defective Filters Marketed for 40 Years: What Philip Morris Never Told Smokers." *Tobacco Control* 11(Suppl. 1): i51–i61.

Platte, Erika. 1988. "Divorce Trends and Patterns in China: Past and Present." *Pacific Affairs* 61(3): 428–445.

Proctor, Robert. 2011. *Golden Holocaust: Origins of the Cigarette Catastrophe and the Case for Abolition.* Berkeley: University of California Press.

Roberts, David. 2015. Opinion: How the US Embassy Tweeted to Clear Beijing's Air. *Wired,* http://www.wired.com/2015/03/opinion-us-embassy-beijing-tweeted -clear-air/.

Rogaski, Ruth. 2004. *Hygienic Modernity: Meanings of Health and Disease in Treaty-Port China.* Berkeley: University of California Press.

Shanghai Tobacco Gazetteer. N.d. Shanghai Local Records Office. http://www.shtong .gov.cn/node2/node2245/node68143/node68151/node68224/node68235 /userobject1ai65762.html.

Stevenson, Terrell, and Robert N. Proctor. 2008. "The Secret and Soul of Marlboro: Phillip Morris and the Origins, Spread, and Denial of Nicotine Freebasing." *American Journal of Public Health* 98(7): 1184–1194.

Zhang, Gangyin. 2007. "Kexue shenghuo: Xiangyan jinghua qi neng xijing "ershou yan" ma?" [Life Sciences: Can Cigarette Purifiers Do Away with Second-Hand Smoke?]. *Science Daily*, June 4.

AFTERWORD

Mei Zhan

We live in a dizzying world of continuously unfolding dreams and nightmares. Every time I have begun writing a coda to this edited volume—a collective exploration into the entanglements of science and technology, the state, the market, and everyday life in contemporary China—a new event would inevitably blow my attempted conclusion wide open. On November 26, 2018, the day before the Second International Summit on Human Genome Editing was to take place in Hong Kong, the Chinese scientist-entrepreneur He Jiankui announced through an exclusive interview with the Associated Press that his research team had produced the first gene-edited babies in the world. He claimed that the babies, a pair of twin girls, would be resistant to the infection of the HIV virus. A "shockwave" would be an understatement to describe the reactions from the scientific communities and the general public in China and beyond. The first set of reactions, appearing through Chinese news outlets on the social media platform *weibo*, celebrated the event as a scientific breakthrough in which Chinese genomic science had finally outstripped its Euro-American counterparts. Could He's experiment be the realization of a long-cherished and elusive "Chinese dream" of national revival?

This celebratory moment, however, was short-lived as voices of concern, criticism, and outrage began to be raised within hours. The first collective response from the scientific community came from 122 prominent Chinese scientists, who signed a joint letter condemning He's experiment as an "unforgivable" breach of scientific ethics that did nothing to advance genomic knowledge or technology (Kolata et al. 2018). The breaking news also turned the International Summit on

Human Genome Editing into a public spectacle. Journalists camped out at the conference, which was convened by the Academy of Sciences of Hong Kong, the Royal Society, the U.S. National Academy of Sciences, and the U.S. National Academy of Medicine. Attendees speculated whether He would show up for his scheduled presentation, discuss his work with leading experts in the field, and answer questions from the audience. He did, on a panel that was live-streamed. So many people tuned in that the popular Chinese video-sharing website, *Bilibili or* "the B Site," temporarily crashed. It was the first time the B Site crashed because of a scientific event.

The summit was only the beginning of mounting inquiries, discoveries, and debates. Genomic scientists, especially those in China, pointed out flaws in He's data and procedure and questioned whether the experiment was a success at all. At the same time, fears of the consequence of genomic science for humankind reverberated throughout China and the world, as some recalled the wondrous and ultimately tragic life of Dolly, the first cloned sheep (Franklin 2007). Further, He's background and networks came under intense scrutiny—where was he trained, who mentored him, and who authorized his research? (It turned out that He received his PhD in biophysics from Rice University.) How did his experiment pass the scrutiny of various levels of ethics reviews? Who funded his research? (He was a faculty member of Southern University of Science and Technology in China, the founder of his own genomics company financed by venture capital, and also purportedly connected with the Putian System, a cluster of private-owned hospitals and clinics that gained fame and notoriety for its aggressive pursuit of market and profit.) At the same time as various state agencies and science organizations—in China and abroad—distanced themselves from He's project, it was revealed that He was once hailed as an exemplar of the Chinese state's 2012 Initiative for Innovation, which would transform China into the global powerhouse in innovation by 2050 (see Amy Zhang, chapter 7).

How can we—as anthropologists and human beings—come to terms with such impossibly complex phenomena and rapidly unfolding events in situated scientific practices and dreams, which make up the worlds we inhabit and at the same time constantly threaten to slip away from our analytical grips? How do existing analytical tools at our disposal both mediate and constrain critical inquiries into the deep enmeshments of science, technology, state, market, everyday civil society, modernity, and the very meaning of being human? Given the intellectual and existential challenges we face, what new forms and frameworks of social inquiry might develop out of a productive conversation between the anthropology of China and science and technology studies (STS)?

The essays in this volume are a valuable exercise and collective effort in initiating and deepening these conversations. Together, they present a compelling argument for why it is critical, at this particular moment, for anthropologists to step in and make their accounts and analyses of science in/of/and China relevant to academic and public discussions and debates. One of the most important points that this collective project makes clear—as articulated by Susan Greenhalgh in the introduction—is this: China is not a place outside of the West where "usual science" proliferates and changes its forms in a non-Western national or cultural context. Rather, the translocal sociohistorical formation and the complex conceptual and institutional interplay of state, market, and technoscience shaping and shaped by post-Mao, postsocialist, and now Xi's authoritarian China *demand* thoughtful and experimental ethnographic engagement on the ground.

This conversation entails a subversion of sorts in the academic echelons. The anthropology of China should not merely enrich the global ethnographic data pool of an STS analytical apparatus that was formed in Euro-American academia and has been interested primarily in Euro-American cases. Early critical social and cultural inquiries into science and technology emerged in the aftermath of the industrial revolution and the Second World War, and under the threat of environmental destruction and nuclear Armageddon—historical events that were part and parcel of Euro-American colonial and post-colonial hegemonies in the world. These events and concerns were grave and pressing, but they should not limit the subject of critical cultural and social studies of science and technology to Euro-American enterprises or interests. In fact, they call for empirical studies and analytical approaches that are both translocal and postcolonial. In recent years, STS scholars have become increasingly interested not only in science, medicine and technology in the Third World and the global South, but also in ways to broaden the analytical and conceptual methods of STS through these ethnographic engagements. These efforts are not always successful, at times they are controversial, but overall they are decidedly worthwhile (see the debates surrounding Law and Lin 2014). In order to seriously investigate how science and technology work conceptually, ideologically, and institutionally in various parts of the world and through uneven translocal fields of power, it is crucial that STS enrich and situate its analytical repertoire through engagements with insights developed in academic fields such as anthropology, which has enthusiastically embraced various theories and methodologies developed by STS scholars (for example, actor-network theory). The essays in this volume offer exactly this kind of engagement by putting discussions of governmentality front and center in exploring science and technology in China; in particular, how scientific and technological concepts, practices, institutions, enterprises and aspirations unfold *through* the shifting translocal nexus of the state, the market, and everyday subject-formation.

Politics and science in China did not go through the exact same steps of separation and collaboration as they did in seventeenth-century England, where, according to Latour (1993) and Shapin and Shaffer (1989), the political philosopher Thomas Hobbes and the chemist-physicist-philosopher Robert Boyle carved up the domains of society and science. As Latour famously argued, Hobbes and Boyle collaborated in creating the Two Great Divides of modernity: first, the Divide between nature and culture, science and society, experiments and politics; second, between the modern European "Us" who (claim to) recognize and abide by the First Divide, and the non-European "Other" who do not (1993). Whereas one of the most daunting and necessary tasks for scholars who study science and technology in the metropoles—and Euro-American contexts more broadly—has been to overcome the Great Divides (especially the first divide) by demonstrating the messy and irreducible hybridity of science and society, the stakes and concerns are slightly different for the "Others" who have never been considered thoroughly "modern" or purified—as the Chinese state, scientists, and ordinary citizens are painfully aware. The entanglements between the Chinese state and Chinese science and technology are not hidden phenomena awaiting discovery by analytical tools developed by STS in Euro-American settings. China's translocally produced power-laden entanglements demand critical and thoughtful studies close to the ground—studies that also push us to develop new analytical foci, tools, and orientations.

The essays in this volume invoke *governmentality* as an analytical point of entry into the enmeshments of science, state, and market and as a way to forge a conversation with topics central to STS literatures. As Greenhalgh points out in the introduction, beginning in the 1990s, attentiveness to governmentality has transformed anthropological studies of Chinese modernity. Moreover, studies of Chinese modernity have enriched discussions of governmentality by noting their limitations, offering critical perspectives, and developing alternative analytics (see, for example, Lydia Liu 1995; Petrus Liu 2016; Rofel 1999). In discussions of post-Mao and postsocialist China in particular, anthropologists have productively used the concept of governmentality to interrogate the workings of biopower in constructing specific bodies, subjects, and populations (Anagnost 1997; Greenhalgh 2008; Zhang 2012) and to examine the ways in which the state reaches deeper into (rather than retreats from) everyday life through market practices (Rofel 2007; Zhang 2012).

The current volume extends and deepens these insights by zeroing in on the ideologies, institutions, and practices of science and technology—not outside of but rather as constitutive of governmentality. The topics they cover articulate pressing concerns and crises in Chinese life and well-being, which traverse epistemological, ontological, and spatiotemporal scales: mental health and domestic life (Ma, Li Zhang, and Kohrman), scientific innovation and aspiration (Song, Mason, and Amy Zhang), science and policy (Ma, Greenhalgh, Lord, and Amy

Zhang), and ecological and environmental health (Lord, Amy Zhang, Kohrman). Furthermore, by examining these phenomena the authors also explore a number of topics and issues of deep interest to discussions in STS—ranging from the politics of quantification, objectivity, infrastructure, interest, to truth claims. In the first chapter of the book, Ma shows that the production of numbers as part of China's infrastructure-building "constitutes not only populations but also communities as objects of governance." (chapter 1). Li Zhang, in her discussion of tenuous relations between "science" and "psy fever" in southwestern China, presents a carefully delineated longer view of "Mr. Science" in the production of Chinese modernity and governmentality (chapter 2). In her examination of China's stem cell research, Priscilla Song showcases problems in establishing the data of "efficacy" for China's aspiring scientists. Her point is not only that data are problematic *in China*, but also that the production of data *is itself* a profoundly fraught process (chapter3). Similarly, in chapter 4 Katherine Mason discusses young Chinese scientists' struggles to produce truthful and objective science and, in doing so, to join the upper echelons of a global scientific community. (If we follow Mason's line of inquiry, He Jiankui's case would be a symptom rather than exception.) Elizabeth Lord follows up in chapter 5 by examining the discrepancy between the Chinese central government's ambitious "green dream" and the difficulty in actually implementing it on the ground. In chapter 6, Greenhalgh shows how the efforts of a multinational corporation (Coca-Cola in this case) to expand in China were re-narrated in Chinese obesity science and policy as discourses of "good science" and "selfless patriotism." In chapter 7 Amy Zhang investigates a set of multispecies relations to ask what counts as "innovation" in relation to more "traditional" forms of knowledges and practices and in so doing pushes us to think whether there is such a thing as indigenous innovation. Matthew Kohrman rounds up the collective inquiry in Chapter 8 by examining how filters—specifically, cigarette filters and air purifiers, as well as the interplay between the two—obstruct and allow flows at the same time, while paying close attention to gendered discourses of personal, domestic, and public spaces.

All these essays are committed to the specificity of science and technology in/of/and China without being reduced to it. And it is this specificity that amplifies the ways in which this volume both resonates with intellectual debates in STS and beyond and offers important insights into particular and effervescent dreams and worlds we inhabit. As Susan Buck-Morss observes, twentieth-century "mass dreamworlds" were constituted through immense material power "transforming the natural world, [and] investing industrially produced objects and built environments with collective, political desire"—often with catastrophic consequences (2002, 2). These dreamworlds and catastrophes, contrary to Buck-Morss' prediction, have not been left behind in the twenty-first century. As I was writing this afterword, a colleague forwarded an article in the *New York Times* entitled "The American Dream Is Alive. In China" (Hernandez and Quoctrung 2018). The authors argue that the economic expansion and upward social mobility China has achieved in the last few decades were unprecedented in history, and that the American dream, though perilous at home, found its articulation in China. None of us is really outside of these dreamworlds (and catastrophes). And this collection of essays has offered us crucial insights into how to live and think critically through them.

References

Anagnost, Ann. 1997. *National Past-Times: Narrative, Representation, and Power in Modern China.* Durham, NC: Duke University Press.

Buck-Morss, Susan. 2002. *Dreamworld and Catastrophe: The Passing of Mass Utopia in East and West.* Cambridge, MA: MIT Press.

Franklin, Sarah. 2007. *Dolly Mixture: The Remaking of Genealogy.* Durham, NC: Duke University Press.

Greenhalgh, Susan. 2008. *Just One Child: Science and Policy in Deng's China.* Berkeley: University of California Press.

Hernandez, Javier C., and Quoctrung Bui. 2018. "The American Dream Is Alive. In China." *New York Times*, November 18. https://www.nytimes.com/interactive/2018/11/18/world/asia/china-social-mobility.html.

Kolata, Gina, Sui-Lee Wee, and Pam Belluck. 2018. "Did a Gene Edit Shape 2 Babies? Experts Tremble." *New York Times*, November 26, A1.

Latour, Bruno. 1993. *We Have Never Been Modern.* Translated by Catherine Porter. Cambridge, MA: Harvard University Press.

Law, John, and Wen-Yuan Lin. 2014. "A Correlative STS: Lessons from a Chinese Medical Practice." *Social Studies of Science* 44(6): 801–824.

Liu, Lydia. 1995. *Translingual Practice: Literature, National Culture, and Translated Modernity-China, 1900–1937.* Stanford, CA: Stanford University Press.

Liu, Petrus. 2015. *Queer Marxism in Two Chinas:* Durham, NC: Duke University Press.

Rofel, Lisa. 1999. *Other Modernities: Gendered Yearnings in China After Socialism.* Berkeley, CA: University of California Press.

Rofel, Lisa. 2007. *Desiring China: Experiments in Neoliberalism, Sexuality, and Public Culture.* Durham, NC: Duke University Press.

Rofel, Lisa, and Sylvia Yanagisako. 2019. *Fabricating Transnational Capitalism: A Collaborative Ethnography of Italian-Chinese Global Fashion (The Lewis Henry Morgan Lectures).* Durham, NC: Duke University Press.

Shapin, Steven, and Simon Schaffer. 1989. *The Leviathan and the Air Pump.* Princeton, NJ: Princeton University Press.

Zhang, Li. 2012. *In Search of Paradise: Middle-Class Living in a Chinese Metropolis.* Ithaca, NY: Cornell University Press.

Contributors

Susan Greenhalgh is the John King and Wilma Cannon Fairbank Research Professor of Chinese Society in the Department of Anthropology at Harvard University. She has worked for many years along the borderline between science and politics/governance and is author of *Just One Child: Science and Policy in Deng's China* (University of California Press, 2008) and *Fat-Talk Nation: The Human Costs of America's War on Fat* (Cornell University Press, 2015), among other titles.

Matthew Kohrman is Associate Professor in Stanford's Department of Anthropology and Faculty Fellow at Stanford's Center for Innovation in Global Health. His research brings anthropological methods to bear on the ways health, culture, and politics are interrelated, as seen in his recently edited volume, *Poisonous Pandas: Chinese Cigarette Manufacturing in Critical Historical Perspectives* (Stanford University Press, 2018).

Elizabeth Lord studies the politics of environmental knowledge production and the effects of China's environmental policies on inequality. She graduated from the Department of Geography at the University of Toronto in 2017, was an An Wang postdoctoral fellow at Harvard's Fairbank Center for Chinese Studies, and is now a postdoctoral fellow at the Institute at Brown for Environment and Society.

Zhiying Ma is Assistant Professor at the School of Social Service Administration, University of Chicago. Her work uses anthropological and historical approaches to explore mental health, disability rights, and population governance in China. She is currently writing a book on the entanglements of psychiatric institutions and family life in China, focusing on the conditions and impact of the recent mental health legal reform.

Katherine A. Mason is Assistant Professor of Anthropology at Brown University, where she conducts research on issues in population health, infectious disease, bioethics, China studies, reproductive health, mental health, and global health. Her first book, *Infectious Change: Reinventing Chinese Public Health after an Epidemic* (Stanford University Press, 2016), examines the ethics and professionalization of public health in China following the 2003 SARS epidemic.

Priscilla Song is Assistant Professor at the Centre for the Humanities and Medicine at the University of Hong Kong. Her work as a medical anthropologist focuses on transnational biomedical technologies in urban China. She is the author of *Biomedical Odysseys: Fetal Cell Experiments from Cyberspace to China* (Princeton University Press, 2017), which was awarded the 2018 Francis L. K. Hsu Book Prize.

Mei Zhan is Associate Professor of Anthropology at the University of California, Irvine. She is the author of *Other-Worldly: Making Chinese Medicine through Transnational Frames* (Duke University Press, 2009). She works on topics including science and technology studies (STS), medical anthropology, China, transnationalism, and globalization. She is currently writing an ethnography entitled "Bring Medicine Back to Life," which examines the emergence of a new kind of "classical medicine" in entrepreneurial China.

Amy Zhang is Assistant Professor of Anthropology at New York University, specializing in environmental anthropology, science and technology studies, and development. She is currently working on a book on state efforts to modernize waste management infrastructures and how such attempts ground and condition the forms and limits of an emerging urban environmental politics in China.

Li Zhang is Professor of Anthropology at the University of California-Davis. She is the author of two award-winning books—*Strangers in the City* (Stanford University Press, 2001) and *In Search of Paradise* (Cornell University Press, 2010)—and the co-editor of *Privatizing China: Socialism from Afar* (Cornell University Press, 2008). Her current project explores how an emerging psychological counseling movement reshapes Chinese people's understandings of selfhood, well-being, and governing.

Index

9 781501 747038